GRAVITY CAUSE EXPLAINED

Combining it with Electromagnetism and Quantum Physics

by Bobby Dee Ticer

DORRANCE
PUBLISHING CO
EST. 1920
PITTSBURGH, PENNSYLVANIA 15238

Dorrance Publishing Co
585 Alpha Drive
Pittsburgh, PA 15238
Visit our website at *www.dorrancebookstore.com*

ISBN: 979-8-8892-5139-2
eISBN: 979-8-8892-5639-7

CONTENTS

ACKNOWLEDGEMENTS

Although I am technically self-educated for not having a college degree, all the free information on Wikipedia and other educational programs on the internet have been invaluable for the continuous development of this book having had countless rewrites. Self-education is allowed by mere effort.

PREFACE

I came up with an idea of how to explain gravity. To compare my idea with other theory, I enrolled in a class at the University of Oregon to learn Einstein's theory of special relativity. I challenged it only to discover it is internally consistent. Shortly thereafter, however, I did not find general relativity to be consistent with special relativity.

The inconsistency pertains to special relativity being a unification of mechanics with electrodynamics and Einstein's failure to include gravity as part of its unification. Constant light speed by special relativity is a limiting condition for matter to neither exceed nor even reach. To the contrary, a present interpretation of general relativity is that it includes a singularity condition of black holes that became, in theory, the origin of our universe as finite and expanding. Even though Stephen Hawking redefined the black hole for it to emit Hawking radiation in order for it to be consistent with the thermodynamic principle of entropy, the assumption of black holes and the singularity are generally accepted by physicists as established, whereas gravity is excluded from standard theory because of various reasons such as a graviton of zero mass being inconsistent with momenta of mass not being explained according to natural cause and effect.

If theory is proven, then it is fact instead of theory. Hawking changed his mind in the 1990s in admitting there is viable evidence supporting the existence of black holes. I agree, but I have discovered an overlooked condition of relativity theory that there is an upper limit being one-half light speed instead light speed.

It only modifies theory insofar as apparent contradictions become explainable paradoxes instead.

Challenging theory is not merely to disprove it; it can also be a means of understanding it. Other means of understanding can result as a simplification of complex ideas as a mixture of simple ones that can easier be understood according to a step-by-step understanding of the historical development of theory.

Although the history of physics provides a means of understanding as a step-by-step process, it generally contains mathematical language too foreign for some of us to understand apart from being taught. There is a multitude of languages consisting of different units of measurement. They contain newtons, farads, amperes, coulombs, ergs and so forth. There are even systems of dimensionless units, such as Plank units. Such language could be foreign to many of us. This book includes a more common language. Units of centimeters, seconds and grams are used along with algebra and geometry without the need of such more complex math as even calculus.

Calculus is useful for the task of formulating theory. A number e, for instance, allows for the avoidance of an infinity problem, dividing by zero, having an undefined result, but dividing by zero can be excluded in the algebraic process as well. If mass M moving at velocity v collides with mass m_0 relatively at rest, then the ratios of velocities and momenta are either infinity are zero. Such ratios need to and can be avoided by calculating results of collision according to conservation of momentum.

Einstein suggested that explaining cause and effect is unnecessary for the mathematical formulation of theory. Learning is achieved by experience, such as history, and by casual explanation, such as how a banana maintains longer life if kept inside a bag. Knowing that light contributes to its aging, just laying bananas inside a dark cupboard is enough to enable them to last longer. However, there are paradoxes in theory instead of contradictions. The clock paradox of special relativity is explainable, but many other paradoxes of theoretical physics have been more difficult to explain, such as how the universe can expand and still maintain gravitational potential the same as that of a black hole. Still, it is explained in this book without the need of more complicated math.

Algebra is numerical math simplified. Its simplicity is about such symbols as letters of the alphabet substituted for numbers. For instance, in place of adding numbers, say 156 and 44, in the manner $56 + 44 = 200$, chosen letters are substituted in the manner $A + B = C$. Instead of multiplying such numbers as 3 and 4 in the manner $3 \times 4 = 12$, symbols are presented in the manner $AB = C$. And $3 \times 3 \times 3 = 3^3 = 27$ can be represented as C^3.

There is also a convenience of symbolic algebra for solving unknowns. For instance, if $5A + 4B = 6C$ and $5A - 4B = 2C$, then the simple steps of addition, subtraction, multiplication, division and substitution are a means of obtaining numerical results. Adding the two equations obtains

$$5A + 4B = 6C$$
$$\underline{5A - 4B = 2C}$$
$$10A + 0 = 8C$$

Hence

$$10A = 8C.$$

Dividing 10 by 8 and reversing order obtains

$$C = 1.25A$$

Substituting the value of C in the original equation obtains

$$5A + 4B = 6C = 7.5A$$

$$4B = 7.5A - 5A = 2.5A$$

$$B = 0.625A$$

All numerical values are thus obtainable by obtaining a numerical value of either A, B or C.

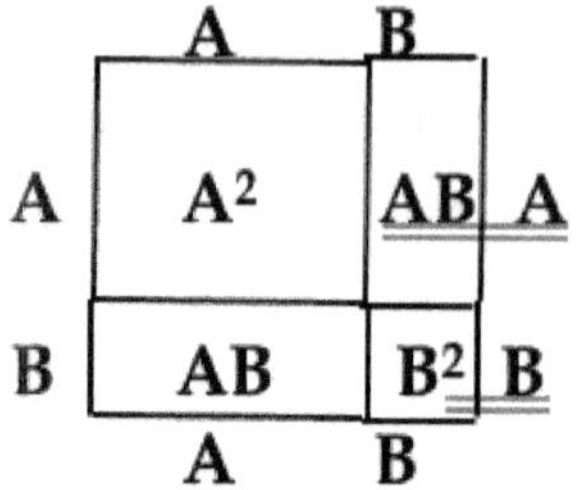

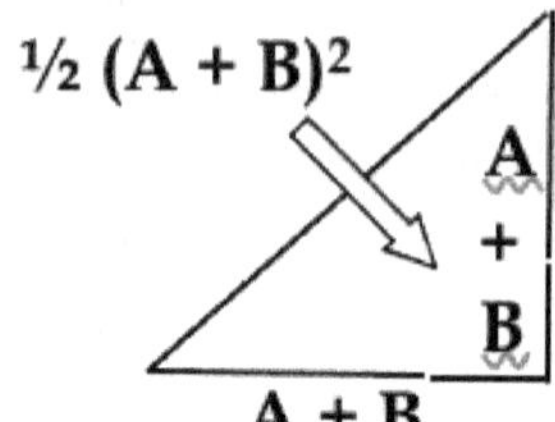

$$(A + B)^2 - 4\left(\frac{1}{2}AB\right) = C^2$$
$$A^2 + 2AB + B^2 - 2AB = C^2$$
$$A^2 + B^2 = C^2$$

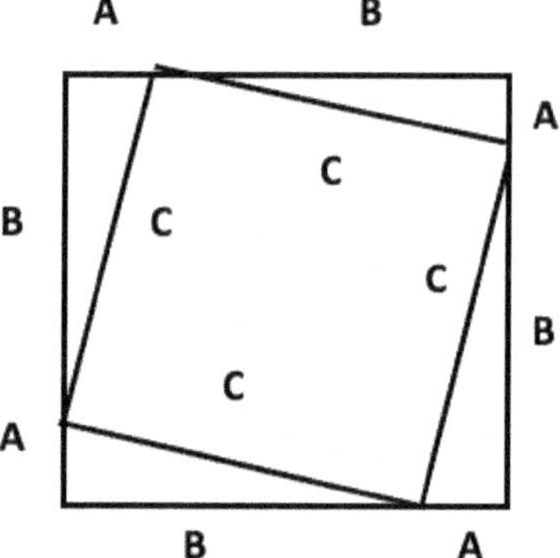

Geometry is also included in this book for more casual explanation. As illustrated above, the Pythagorean theorem is an example of an invariant, which is an essential part of relativity theory. Explanation of it according to algebra is that the hypotenuse length C of a right triangle equals the square root of each perpendicular side length squared and then added in the manner $C^2 = A^2 + B^2$.

A square inscribed within a larger one is that of a larger square $(A + B)^2$ whereby its side lengths A + B are also side lengths of four right triangles. Being that their hypotenuses of each of the four right triangles equal the sum of each other side squared, being the same as that of the inner square, the total area of the inner square is only that of the outer square minus the area of the four right triangles. Hence, instead of $C^2 = (A + B)^2 = A^2 + 2AB + B^2$, total area 2AB of the four right triangles is subtracted out in compliance with Pythagorean theorem formula.

Note: Simple steps of addition, subtraction, multiplication and substitution are here implied, as left for the reader to apply. However, the math is merely for verification instead of understanding of theory. It can be skipped over in this book if understanding is itself a means of verification.

1

INTRODUCTION

Rene Descartes philosophized, "I think, therefore I am. If I am deceived of my existence, then I must at least be the Deceiver." He therefore confirmed his existence by way of his own awareness of himself.

Am I alone?

Resistance to thought and action is testimony to an extended existence of an objective world in which we live. I am not alone. We therefore exist.

I witness existence, but I do not seem to know how either it or a state of consciousness is possible. I merely assume we are parts of creation by either a Supreme Being or from what merely exist.

Can there be existence without our awareness of it? No matter. Never mind. The mind-matter duality is of no concern. This book is merely about the physical world in which we live inasmuch as material existence is comprised of a substance of some sort that varies in shape, size, density and so forth. In this regard, philosophers have attempted to explain everything existing in the world as part of a primary substance that became referred to, according to various spelling, as either æther, aether or ether.

Suppose all physical reality is indeed comprised of ether. What, then, are its properties? How does it, for instance, separate from and recombine with itself in creating effects of reality? A substance being primary implies it has no internal mechanism to bond with other primary substance except for it interacting with its other parts. If such interaction is relative motion, then the action must somehow be elastic to maintain itself in that action between itself would otherwise result in a loss of relative motion.

Another likely condition for primary substance is that it needs to be of infinite content, even if only to partially fill infinite space instead of all of it, in order for it to change direction by means of elastic collision instead of spreading apart without a means to reverse direction to again interact with itself. (To the contrary, however, our universe appears to be finite and possibly expanding.) The question then comes to mind as to whether space is an ethereal plenum or a partial vacuum partially filled with ether.

Descartes, along with other philosophers, assumed space consists of a plenum. How, then, do various densities of matter exist if ether is everywhere identical in composition?

A plenum does not even have wiggle room for wave action to occur in the manner sound waves are compressed states of air molecules. The ether as a plenum is thus contrary to the air medium for sound.

It is no wonder that early Greeks deep in thought regarded our world as illusionary. As for the plenum, some of them considered only circular motion in it is possible. The sensation of free motion in various directions is allowed by the complexity of circular paths for our witness of various effects.

Descartes also assumed all motion is circular as a complex system of vortices of diverse sizes and rotational speeds, and he further assumed total motion is conserved in manner of being allowed by an exchange of motion between various actions of ether. As for the complexity of motion, endless possibilities exist by means of vortices moving inside other vortices for various relationships. A number of like vortices in one region of space can thus be relatively more or less dense than other vortices in another region of space. Our perception of the world thus comes about as an exchange of motion between vortices even though primary substance remains everywhere in space the same.

The source of all creation could be creation from what already exists ad infinitum, and whatever existence we are aware of could be only part of many realities. Other parts of existence could, in effect, be invisible to us.

The ether is not presently accepted as scientific because of being invisible if it does exist, but modern quantum mechanics is somewhat similar to it. In place of the ether, there is vacuum space that need not be empty, consisting instead of a virtual field of virtual particles detectable according to a condition of probability that is mathematically predictable and thus indirectly confirmable according to observation. However, indirect perception of the atomic world of quantum interaction is different from that of the natural world of direct observation. The main difference is with regard to change. Interaction within the atomic world is of discrete units of energy whereas change of particle interaction, either by relative motion or gravity, is of a continuous nature.

This book merely provides explanation of continuous effects of relative motion and gravity in manner consistent with the indirect quantum effects. The explanations are a step-by-step process in view of the historical development of theory for there to be no contradiction of quantum mechanics and relativity theory.

In chapter 2, titled From Aristotle to Galileo, the early history of Aristotelian physics was followed by a step-by-step process to the discovery of laws of nature. Light, according to the early physics, provided earthly substance with the energy needed to maintain motion. In response, impetus theory evolved whereby the material substance of matter maintains an innate ability to sustain motion until impeded by some other force consistent with that of other matter. From two different viewpoints, space was either a plenum or a partial vacuum. Copernicus and Kepler promoted the latter with the heliocentric theory of planetary motion around the sun, and Galileo further explained natural laws of motion in view of a vacuum state, but the plenum was defended by such other philosophers as Gassendi and Descartes. They regarded the existence of an internal mechanism as essential for giving rise to the properties of the observable world.

Although the laws of nature have by themselves provided much understanding of nature, there is still the internal mechanism of reality as inclusive to its understanding. A ratio of mass to volume space of virtual particles, for instance, relates to chemistry and biology, and the ratio of mass to volume space of some virtual particles led to such other ratios as that of the proton and electron.

In chapter 3, titled Newtonian Mechanics, it is explained according to Kepler's laws of planetary motion and Galilean relative motion, but not as a complete theory by itself. Newton, himself, was dissatisfied that he was only able to explain gravity according to an action at a distance principle. He also assumed ideal conditions of absolute space and absolute time that were later modified by Einstein. His mechanics also lacked a definition of energy that was later theorized according to the laws of thermodynamics. Nonetheless, even though Newtonian Mechanics was to be modified by Einstein for it to comply with the relativity of spacetime, it is still an integral part of theory as needed for a more complete understanding of it.

In chapter 4, titled Kinematics Atoms and Electrodynamics, they are explained in the historical context of Boyle's law discovered with the aid of several inventions that gave rise to explaining heat and temperature in view of the kinematics of relative motion, which eventually led to atomic theory and the laws of thermodynamics. The second law is known as entropy. It is significant in describing the determination of interaction according to various equilibrium states. It soon was used to verify a Fourth Power law of black body radiation and to become part of the development of quantum theory.

In chapter 5, titled From Wave Theory to Relativity, it is about a wave theory of light in contradistinction to the partial vacuum of space. The theory developed around 1900 to explain such properties as diffraction, but its requirement of light waves being transverse waves required explanation itself as to how transverse waves can exist in a three-dimensional medium in contrast to surface waves or waves along a rope. A solution was provided by Maxwell's electromagnetic theory as light waves being a continuation of magnetic fields induced by electric currents in free space. The speed of light was included in the theory as a constant with a determined value nearly at which it is presently determined. Constant speed through the ether was assumed, but the state of the ether remained questionable as to whether it was dragged along with mass or was an absolute state not influenced by the presence of mass. Experiment indicated that light speed measures the same regardless of the relative motion of the system by which it is measured. An explanation was given by Lorentz whereby contraction of material length in the direction of relative motion and clocks being slowed by their motion in the ether contributed to the measure of light speed as constant.

Lorentz did not conclude that the ether is empirically indeterminable. He left that conclusion to Einstein, who further proclaimed ether is thereby of no use for the formulation of theory. It thereby became non-existent in order for physicists to describe and explain only the natural world according to mathematical formulation, such as what later became the probability of observational facts, but Einstein also suggested the ether could provide a deeper understanding of the nature of the universe in a way that he did not apparently pursue.

In chapter 6, titled Simple Spacetime Relativity, the relativity of space, time and motion is shown to be an internally consistent theory according to the postulates of covariance and constant light speed. Consistency of theory includes the principle of simultaneity, relativistic transformations, the addition of velocities theorem, constant speed change and the clock paradox, all of which is verifiable according to algebra.

In chapter 7, titled Mass-Energy Dynamics, special relativity is shown consistent with such conservation laws of momentum and energy. They differ somewhat from Newtonian mechanics in that elastic collision is modified to only be partially elastic in order for mass itself to neither reach nor exceed light speed. Mass increasing speed from collision with either other mass or light energy thus retains part of the other mass or energy (as mass-momentum) as the inelastic part of collision in order to comply with conservation of both momentum and energy. However, a paradox to be resolved is how internal mass-energy relates to massless photons of light energy or massless gravitons of gravity.

Chapter 8, titled The Relativity of Gravity, is more complex. Einstein attempted to generalize the principle of covariance to include gravity, but it is complicated by the inhomogeneous nature of gravity, such that gravity is described according to spacetime curvature due to the presence of mass. An energy-momentum tensor is used to describe spacetime curvature, but there are nonetheless conditions of relativistic effects of general relativity that are analogous to those of special relativity. Gravity is, in effect, another form of acceleration whereby light speed varies according to gravitational field strength.

Relativistic effects of acceleration are analogous to the clock paradox such that homogeneous conditions of gravity as distance and time increments are analogous to inertial motion whereby a Schwarzschild metric of general relativity links to a Lorentz metric of special relativity. It appears Einstein did not consider the link in his failure to formulate a unified field theory. It could have been that he preferred to concentrate more on explaining how the presence of mass results in spacetime curvature. Although theories can be mathematically interrelated, predicted effects yet to be discovered and various interpretations of discovered effects have mainly prevented the unification of all theory.

Explaining spacetime curvature in more detail results in slight modification of both special and general relativity. Spacetime curvature is according to local perspective. Because clocks are slowed by gravity, light speed needs to be slower as well. More complete explanation of relativistic effects indicates an upper limit of one-half light speed instead of light speed. The modification allows apparent contradictions to be explainable paradoxes.

A controversial issue of general relativity is regarding the universe as finite and expanding. Einstein considered a finite and static universe, and he inserted a cosmological constant in his field equations in order for it to be a counter force not allowing gravity to continually condense mass into a smaller volume of space, but Friedmann pointed out to Einstein that a static universe was not stable according to the general field equations, even with Einstein's insertion of a cosmological constant. Moreover, with the discovery of a redshift in the spectrum of more distant starlight, it indicated to Friedmann and other physicists that the universe is expanding instead. A metric was proposed by Friedmann and the other physicists to describe it, in part, according to general relativity. Apart from relativity theory is the origin of big bang being a tiny speck in comparison to the smallest atomic element.

The universe, according to big bang, is theorized to have evolved from an extremely dense and tiny speck of energy source in contrast to the black hole condition of general relativity that only applies to direct observation of our observable reality. However, covariance is also part of the theory insofar as observers at the edge of the universe perceive themselves to be at its center instead. A link between general and special relativity provides interpretation whereby black holes exist within black holes in a way of allowing our universe to merely be one within an infinite number of them.

In chapter 9, titled Quantum Origins, a Planck constant is explained in the historical context of theorizing the natures of atomic energy as quantum states of existence in contrast to gradual change in effect according to relative motion and gravity.

In chapter 10, titled Quantum Particle Physics, Bohr's quantum theory of the atom is explained. Its development by Bohr includes Newtonian mechanics.

Further inclusion is the photoelectric effect explained by Einstein and a possible distinction between quantum and gradual change conditions by other physicists. Bohr had assumed the electron's orbit around the atomic nucleus is a circular path. Other physicist proposed it is elliptical instead in allowing for slight differences in effect for being compatible with how gradual change occurs in reality, and for a possible explanation of how the fine structure constant relates to a speed about 137 times less than light speed.

Chapter 11, titled Quantum Wave Mechanics, is a wave interpretation of the Planck constant proposed by Schrodinger. A wave-particle duality was previously proposed in the mid-nineteenth century by William Hamilton. Louis de Broglie proposed a wave-particle duality theory interpreting quantum effects consistent with relativity theory. It is somewhat speculative. Schrodinger's wave mechanics theory was developed according to musical harmony. Although it included significant predictions, physicists preferring particle theory explanation reinterpreted Schrodinger's wave equations as probability conditions.

Chapter 12, titled Æther Cosmology, is about Harold Aspden's development of theory in favor of recognizing the ethereal medium as part of its inclusion. It combines a Thomson electromagnetic formula of internal mass-energy, previous to Einstein's formula of it, with a Coulomb electrostatic formula relating to kinetic energy. Further formula equates how the proton in the nucleus is a combination of smaller particles. His theoretical development further includes how protons combine to emit gravitational effects that are mathematically consistent with the value of Newton's gravitational constant G.

Chapter 13, title The Relativity of Hubble Cosmology, distinguishes the difference in Tired Light and Big Bang theories. Big Bang assumes the universe is finite and expanding whereas Tired Light explains the Hubble constant as a decrease in light energy while propagating through the medium of space. Either way, the Hubble constant further equates the ratio of gravitational and electromagnetic force according to its own value per light speed at a distance equal to the nuclear diameter of the hydrogen atom.

Chapter 14, titled Mass Motion Attraction and Equilibrium States, is a more general explanation of the cause of gravity along with electrostatic attraction and so forth. Overall, the unification of theory is not that quantum electromagnetic effects equate with gravity; they are explained instead only as different but complementary states of equilibrium. Gravitational potential increases with increase in mass whereas electrostatic potential decreases with increase in mass. They do not directly equate. Unification only pertains to how they mathematically interrelate. They equate indirectly by means of an astronomical recycling process according to the Hubble constant and equilibrium states in compliance with such principles as entropy and the relativity of covariance. Mass is assumed to be formed according to how it either reflects light energy or allows it to pass through as invisible wave action. A Pauli exclusion principle is referred to regarding how waves can superimpose whereas different matter cannot occupy the same space. The superimposing allows energy to escape in manner consistent with entropy for vacuum effect to occur according to different conditions of equilibrium regarding the nature of the quantum and so forth.

There have been explanations of gravity along with unification of theory. Their non-acceptance by a Standard model is likely because of having no experimental verification of the different possibilities. Even though gravity cause along with unification is explained in this book, it is not proven. It is mainly for understanding and possible future contribution to theory.

There have been many possible unified theory proposals. String theory is an example. It is complex and only explained here inasmuch as it describes various effects according to how strings form into various dimensions of space. It is still being explored with regard to discovering predictions and the detection of unexplained effects. It is now a main part of observational research and theory formation, but it is only speculated later on in this book with regard to how it could relate to the circular condition of the plenum.

2

FROM ARISTOTLE TO GALILEO

Circular motion was conceived in ancient times as divine, as evident of stars orbiting in the heavenly sky above earthly chaos. Earthly substance was considered the center of the universe below heavenly stars.

As to explain the primary source of motion itself, Aristotle (384–322 BC) proposed an Unmoved Mover is its provider. What evolved from this proposal was a theory of emanation. Later, for instance, Robert Grosseteste (1168–1253) and Saint Bonaventure (1217–1274) proposed God first created lux, a corporeal form of substance that duplicates itself indefinitely. Motion consists as the duplication of form as waves of energy moving in all directions. Lux constitutes the material form of substance by reflecting lumen, which is light. Light is also how God mediates between souls and bodies. As nature takes its course, it is not alienated from God, as He intervenes by emanating light from within.

Light, according to Aristotelian physics, is thus the essential source of motion, as it is distinguished from earthly substance having no inclination whatsoever to move without assistance. This theory was challenged by John Philoponus (about 490–570 AD) in asserting material substance is inclined to remain in motion without assistance. This idea identifies with inertia whereby a state of non-acceleration, mass in relative motion or at rest, maintains unless it is changed by means of an external force, such as gravity or a collision from other mass in relative motion. However, even though it was an insightful idea for advancing theory, it was instead rejected by theologians with more influence in favor of the Aristotelian doctrine.

Light, in modern physics, is still an essential part of mass. The internal energy of mass is $E = mc^2$. The difference of it from Aristotelian physics is the inert aspect of mass allows relative motion to continue, but light is still an internal source of change, and there is further distinction of matter and light to consider. Matter varies in speed from interacting with light or other matter whereas light only varies in speed if moving through material media, such as water or air, or in a gravitational field. If matter changes speed by its interaction with light, both momentum and energy are conserved according to either classical mechanics or relativity theory. (As such, it is to be mathematically verified in another chapter of this book.)

Impetus and Inertia

A modification of Aristotelian physics did not fall on deaf ears outside of Europe. As during the golden age of Muslim academic culture in Persia, Avicenna (980–1037) concluded motion is an inclination transferred from the thrower that does not cease if it occurs within a vacuum. He obviously realized a decrease in motion requires a resistance to it, such as by the presence of air. This reiteration of the Philoponus position identifies with a modern concept of inertia and momentum in view of empty space, but the plenum was still embedded in general thought. Philoponus and Avicenna, for instance, both conceded that the power of motion given to an object to move through a medium would eventually be used up. Does, then, the light moving through the gravitational fields of the universe surrender its energy to them? It is considered in this book that it does according to how light and gravity interrelate. Light propagation within a gravitational field is slower in proportion to the strength of the field.

Saint Thomas Aquinas (1223–1274) and Francis de Marchia (b. around 1285–d. after 1344) also accepted the idea of motion of matter maintaining until it is impeded by the presence of another object, or by gravity, but the idea of matter having innate ability to propel itself forward indefinitely with no additional assistance was proposed by Jean Buridan (1295–d. after 1358).

Buridan proposed that motion given to an object from another object is sustained by the object until passed onto another object. He named this inherent property of motion impetus. He did not identify impetus with the energy of light, but he did offer a biblical justification for it according to his interpretation of Genesis, stating God rested on the Sabbath after He created the world in six days. Because God rests, He allows His creation to sustain motion such that no longer is there any need for Him to replenish it.

Impetus theory is identical to the modern concept of relative motion except that Buridan referred to rest as distinct from motion to allow for an underlying medium such as the ether for a state of absolute rest for it to be distinguishable from the relative motion of all matter moving through it. His theory remains consistent with the modern mechanical interpretation of motion insofar as Buridan even stated impetus is proportional to weight times speed. A heavier object or a faster one thus has more impetus, which compares with more momentum in view of modern terminology. Since gravity provides impetus to increase motion towards earthly mass and earthly mass gives up impetus to escape from earthly mass, a cannonball falling through a hole through Earth is increased in impetus on the way down to the center of Earth the same as the cannonball gives up impetus to move up an equal distance to the surface at the other end. Such analysis later exemplified such periodic motion as the free swing of a pendulum and of oscillatory motion in general. The latter was to be theoretically developed in the seventeenth century, but the concept of impetus was interpreted differently by other thinkers in the fourteenth century.

Nicole Oresme (1320–1382) maintained impetus is the temporal quality used up in motion by the inertness of earthly substance tending toward its natural place of rest, the Earth, as Aristotle had contended, and as evident of objects losing motion by falling to the ground. He further distinguished between an impetus given to the motion of the heavenly stars and impetus given to the violent and accidental motion of earthly events. However, he argued, contrary to Aristotle, that it cannot cause an object to accelerate to an infinite speed even in the vacuum of space because the impetus is spent during motion, which would be correct if impetus were to be identified as the cause of acceleration.

Oresme further considered Archimedes' principle of the lever whereby the position of a heavier object is placed nearer to the fulcrum for balance. In interpreting this principle as applying to the cosmos, Oresme referred to the Aristotelian idea that earthly substance tends toward the center of the world where Earth rests. However, the moon, sun and other celestial bodies moving about indicated to Oresme the center of gravity could shift. Thus, it is possible Earth can slightly shift in position as well. He further reasoned, however, that Earth's movement is not proven one way or the other. He further proposed the criteria two equal hypotheses should be the merit of simplicity, as Copernicus later advocated, but Oresme accepted a stationary earth in support of the common interpretation of the Bible at the time.

The Copernican Revolution

While impetus theory developed at Paris, France, more eloquent ways of describing nature developed at Merton College in Oxford, England. Such scholars as John Dumbleton, Richard Swinehead and Thomas Bradwardine proposed an abstract system of degrees and latitudes for analyzing qualities of nature, such as hot and cold, and various forms of motion according to quantity.

The intent of these Oxford scholars was for providing a mathematical description of processes rather than to claim their abstract calculations were actual laws of nature. They, nonetheless, provided the quantitative means for mathematically analyzing results of experiment in arriving at such concepts as constant acceleration and instantaneous velocity. This is particularly true with Domingo DeSoto (1494–1560) applying the calculating technique of the impetus theories developed at Paris to calculate a mathematical idea put forth by Oresme and others that increased speed of falling objects is due to constant increase in impetus.

After this development, Nicholas Copernicus (1473–1543) proposed a heliocentric scheme whereby planets, including Earth, revolve around the sun. His scheme was rejected by authoritative rule, but it was later to be advanced by Johannes Kepler (1571–1630).

Aristarchus of Samos (b. around 310 BC–d. about 230 BC) advocated a heliocentric system of planets, including Earth, revolving around the sun. However, his scheme was overcome by the lack of knowledge regarding the condition of parallax.

Parallax is the apparent change in position of an object caused by the observer moving here and there. The positional change appears greater for closer objects. Nearby scenery, for instance, changes rapidly for passengers inside a moving automobile looking out the window, whereas a more gradual change in relative position occurs of more distant mountains. The change in the relative position of the sun appears not to change except for Earth rotating for day and night to occur. Earth revolves around the sun as well for a change of seasons. Such slight change can be interpreted as parallax of either Earth's or the sun's motion. Description of the latter is sometimes more complicated. A complex epicyclic system, for instance, was used for describing circular paths of planets and the sun revolving around Earth.

Copernicus did not consider parallax. Apart from it, observations of astronomers provided more and more data on the relatively closer positions of planets in our own solar system such that the scheme of circles within circles for planets moving around Earth was more complicated. Copernicus thus proposed the much simpler scheme of Earth circling the sun. He also asserted no internal effects of Earth moving through space are detectable since all relative parts move uniformly, which implies a principle of relative motion preceding its proposal by such other theorists as Galileo and Einstein. Galileo stated the laws of physics are relatively the same to all systems regarding velocities of inertial motion. It preceded the development of both Newtonian mechanics and Einstein's relativity theory.

Publication of Copernicus's book *De Revolutionbus* occurred the same year of his death, but the book was outlawed along with later works by any of its defenders, including Galileo. Giordino Bruno (1548-1600) advocated, for instance, an entire world full of solar systems, and he further speculated stars move relative to each other but that they are too distant from us for detection of their relative motion. However, he was burned at the stake for his outspokenness, whereas Galileo was merely sentenced to confinement for his defiance of the order.

Copernicus had nonetheless set forth a revolution in thought. Readers in Italy and elsewhere accepted *De Revolutionbus*. Simon Steven (1548–1620) of the Netherlands supported the heliocentric system with his book *De Hemelop,* published as early as 1608. However, the Copernican system was not faultless, as celestial data compiled by astronomers indicated planetary motion was not of true circles. As Copernicus revered the circle as a divine principle, his scheme included thirty-four epicycles. This weakened his argument of simplicity. A Copernican revolution had nonetheless begun. Ironically, Johannes Kepler, a so-called mystic who claimed to listen with a sensitive ear to the musical harmony of planets in motion, including Earth, was to defend it.

Kepler's Celestial Scheme

Kepler (1571–1630) was proficient in mathematics. Such skill enabled him to become an assistant to Tycho Brahe's (1546–1601) in plotting data of celestial movement. Brahe had rejected the heliocentric system in favor of a stationary Earth mainly because he could not detect parallax of the stars. He regarded Earth material as an inert condition of nature. However, he did accept a heliocentric system as applying for all the planets other than Earth revolving around the sun. As he maintained the sun and moon only revolve about Earth, it would only be another step to include Earth as revolving around the sun.

Kepler studied the data compiled by Brahe and, in 1609, proposed three empirical laws to describe it: (1) planets move in elliptical paths, (2) an equal area in equal time is swept between the sun and a planet, and (3) orbital periods squared in ratio to the cube of the planet's average distance from the sun are the same for every planet.

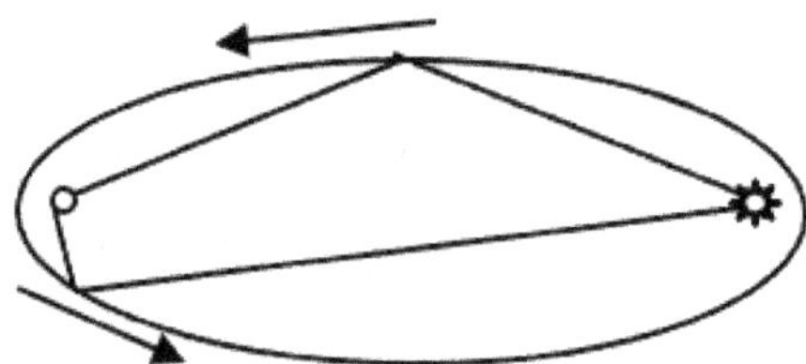

In view of the first law, an ellipse differs from a circle in that it has two foci in place of a center. A property of an ellipse is a total distance between any two straight lines connecting the two foci to anywhere on the perimeter of the ellipse is always the same. As illustrated above, consider a pencil maintains a taut string between two locations for drawing an ellipse. The length between the two loci determine the eccentricity of the ellipse. The total length of the two upper lines in the above illustration also equals the total length of the two lower lines, and the average of length between either of the two respective lines is also the same as the radius of a circle. If two foci separate from the center to the edge of the circle, for instance, the circle itself becomes its diameter. A semi major axis of the ellipse, as the average length of two foci lines, thus equals the radius of the otherwise circle.

The second law refers to the distance between the sun and a planet sweeping the same area per time. The focus position of the sun is more massive than focus positions of planets for the planets to orbit around the sun. The second law states the area swept per time between the sun and planet by the planet's orbital motion is the product of the distance of the planet from the sun and the speed of the planet at its distance from the sun. Planets thus move faster when closer to the sun for them to maintain the same amount of area per time.

Note: An ellipse with both its two foci in the center is a circle. In any case, the average length of the two foci equals the radius of a circle, and orbital speed of a circle is the average orbital speed of the ellipse. Because twice as much radius has four times more area for a perimeter of twice distance, a twice speed increase at half radius takes one-fourth as long to sweep four times more area, being the same area swept per time.

The ellipses are not the same for each planet, but Kepler determined a common property as his third law. By it, a planet's orbital period squared in ratio to the cube of the planet's semimajor axis as its average distance from the sun is the same for all planets. For simplicity of calculation, consider an Earth's orbital period to be one year. If its average distance from the sun is also determined as one unit, chosen as such, its orbital period squared in ratio to its semi major axis cubed is one numerical unit, which allows for a simpler means of comparing the ratios of the other planets with Earth's.

The average orbital distance is that of the semimajor axis. The orbital period is orbital distance per orbital speed. Both distance and speed of orbit determine the orbital period. A planet with a semimajor axis that is twice that of Earth's has twice the orbital distance while moving at a slower rate. If slower by the square root of 2, its orbital period is 2 (according to the longer distance) multiplied by the square root of 2 (according to the slower speed). A period squared is the square of 2 multiplied by the square of the square root of 2: $(2\sqrt{2})^2 = 8$. The twice more distance of a semi major axis cubed calculates as $2^3 = 8$, and 8 divided by 8 is also unity, the same ratio of Earth's period squared and semi major axis cubed as unity.

It is possible Kepler could have had insight of the laws of nature, but they were developed in a progressive manner by other legends of history, as by Galileo and Newton. According to Newton's inverse square law of gravity, for instance, the average orbital speed v relates per average orbital distance r as centripetal acceleration v^2/r. Orbital speed around the same mass at twice radius r decreases orbital speed v by the square root of ½ for its orbital period at twice orbital distances 8 times as much, being consistent with Kepler's third law.

Newton's inverse square law of gravity also connects with Kepler's second law in terms of orbital speed. By Newton's formula, Gm/v_2 for orbital speed to be v. At one-half r, v_2 becomes $(2v)_2$, being consistent with Kepler's second law.

Terrestrial Mechanics

Physics refers to the laws of motion as mechanics. Complementing the celestial mechanics of the planets is terrestrial mechanics. Its development also connects with gravity determining the nature of bodies falling to Earth.

An experiment on gravity was done in the third century BC by Strato in determining difference in sounds of falling bodies hitting the ground from different heights. He thus surmised an increase in speed occurs during the fall as evident of the louder sound of impact from an object's fall from a greater height.

An experiment in the thirteenth century was an attempt by Jordanus de Nemore (1225–1260) to distinguish weights of objects according to their angles of decent while they slide along planes. His theory of positional gravity and component forces considered work as in relation to the position of a level apart from where it balances in a state of equilibrium.

Leonardo da Vinci (1452–1519), whose work did not all survive except for his notes, attempted to determine if the gravitational fall of an object is directly towards the center of Earth. He dropped two heavy objects from a tower in a failed attempt to find a decreased distance of separation.

Da Vinci failed to determine the direction of fall because the change in distance is too minute to detect with the instruments he had available, but he discovered instead a pyramidal increase in the speed of fall occurs in equal intervals of time in analogy to counting stairs. However, from his notes, which might not necessarily reflect what he actually concluded, he incorrectly stated the distance of fall is proportional to time instead of time-squared.

Oresme, Galileo and others correctly determined the distance of fall is proportional to its time squared, but da Vinci might have only erred in taking notes of his findings. To his credit, da Vinci seems to have been wise to the ways of nature. For instance, in anticipation of Newton's third law of motion, whereby force and the resistance to force are mutually the same, he suggested air and water counteract with the same amount of force.

Demonstratio, published in 1552 by Giovani Battista Benedetti (1530–1590), was a book that attempted to determine the nature of falling bodies analytically. Benedetti first assumed bodies of different weight fall at the same rate if they are equal in density and composed of the same material. His proposal was contrary to Aristotle's doctrine that a heavier body falls faster than does a lighter one, but in a revised edition, published in 1554, he changed his position to that of bodies of the same material but of different size do not fall at the same rate.

Perhaps critics influenced Benedetti to change his position, or perhaps geometrical considerations were apparent. If an object divides into two or more parts, for instance, more inside becomes part of the outside, which is also more exposed to atmospheric conditions. Since only the surface area changes instead of total volume of all its parts, this means the mathematical ratio of volume to its surface area is relatively according to its size. Smaller objects thus encounter more friction per surface area to volume or weight, which causes them to move more slowly through the atmosphere.

All bodies of mass falling at the same rate could indicate empty space apart from the atmosphere. However, there could still be a medium of space that mass moves freely through by means of adjustment. It would be similar to how an impulse of momentum moves a solid material as wave action. If the medium is of identical particles, then momentum of mass is simply transferred from one mass to another. However, as it were, for supposedly not knowing the true nature of mass, Galileo Galilei (1564–1642) theorized instead that all bodies–regardless of their size, weight or material they are comprised of–fall, in vacuum, at the same rate.

The abstract analysis of motion at Merton College likely guided Galileo rather than the works of Benedetti. In any case, the hypothesis Galileo put forth provided a means of testing whether the laws of motion typify objects moving through empty space as would seem necessary for the planets in the Copernican system to move unopposed by a medium. Galileo therefore experimented with objects moving along planes and took notice of the free swing of pendulums to discover an appreciable amount of reduced friction tends to allow motion to maintain. He thus postulated the first two laws of motion that Newton would later formulate in his system of mechanics with regard to the inertia of continual motion and its acceleration as a change in either speed or direction, or both.

Galileo made such other contributions as to construct telescopes from the discovery of magnifying glasses by others. He even examined tension of strings in regard to music. In order to test the equality of fall between masses, it is likely Galileo performed such experiments as is alleged of him dropping objects off the leaning tower of Pisa.

Galileo also noted how difficult it is to obtain accurate results from such difficult situations. He seemed to take other results of experiment for granted as well, but as indirectly confirmed by other experiment. For instance, Pierre Gassendi (1592–1665) had directed an experiment to be performed on a moving ship at sea to find out if an object follows a straight course of the ship while falling to the foot of the mast. His own experiments having already verified his laws of motion, Galileo confidently asserted the object would fall the same as if the ship did not move.

To his credit, Galileo truly deserves the acclaim of pioneering the modern approach of establishing laws of nature according to observational facts rather than by ontology or abstract concepts of intuition. He is given credit for developing the procedure of developing theory according to experimental results.

Weighing the Debate

The idea all of space is filled with an undetectable medium is contrary to the empirical approach, but the new discoveries do not prove space needs to be empty for mass to move through it. It only affirms objects move as though empty space is before them instead of it actually existing before them. Moreover, the new mechanics is not a complete explanation of reality. It does not explain, for instance, how it is possible for corporeal matter to attain and maintain a form in the manner it does. How, for instance, are elastic collisions possible without an underlying force to maintain material form?

Such questions were still being pondered. Francis Bacon (1561–1626), for instance, questioned such concepts as a vacuum state with regard to the nature of matter. If matter consists of individual atoms, then how are they kept intact? He concluded atoms need to somehow possess inner qualities by means of some intangible spirits arising from a medium of some sort to provide cohesion and form. He also advocated a primary role of science is to describe nature according to how it is observed.

Another philosopher who did uphold the atomic theory as an internal mechanism was Gassendi. He offered an atomic theory in view of primary and secondary effects. Secondary effects are of inertia and motion. Inertia is necessary to resist penetration and to change the motion of other atoms by means of direct contact. The atoms sometimes combine to produce various effects, such as for our observable world to be comprised of the secondary effects arising from a primary source that produces the form and cohesion of the secondary effects. The primary source is visibly indeterminable by us even though it gives rise to the secondary effects that actually constitute the natural world of observation.

A philosopher who went so far as to advocate a plenum in view of the concepts of relative motion and inertia was Rene Descartes (1596–1650). He was well aware of the implications of these new concepts, but he opposed the vacuum state. He thus undertook the task of explaining relative motion and such effects as gravity in view of a plenum.

Since motion through a plenum is by reason necessarily circular, the Cartesian universe contains vortexes that differ in size and rotational speed. Exchanges occur by smaller vortexes moving at different speeds in determining various effects, and the visibility of the world is determined by the size of our nerves extending from our brains. Our seeing as humans is, in fact, dependent on how our brains can comprehend all the many images our eyes allow it to focus on.

Descartes further postulated conservation of motion, which is consistent with conservation of momentum in that a collision between two masses results in a change in speed of the greater mass being less than a change in speed of the lesser mass. Twice an increase in speed of one half as much mass nullifies a half decrease in speed of the twice mass to thus conserve total motion. However, even though it complies with conservation of momentum, difference results in kinetic energy because of it relating to speed squared. By conservation of motion, regarding inelastic collision for simplicity, consider twice unit mass m_2 moving at unit speed v_1 collides with unit mass m_1 relatively at rest. For conservation of momentum, the resulting speed of m_e needs to be 2/3 unity for the numerical value to remain at $3(2/3) = 2$. However, for conservation of kinetic energy, it changes from $m_2(v_1)^2$ of the numerical value 2 to $(3)(4/9) = 4/3$.

With regard to inelastic collision, motion can be conserved with the creation of heat either as the internal motion of molecules or as radiation, as is light. Kinetic energy does not appear conserved by inelastic collision except for it being internalized in some other form, such as heat.

What continual motion in particular directions indicates is separation of an isolated system of finite size unless there is something beyond it to reverse its internal motion. It thus seems that interaction within a finite universe needs to somehow extend outward to infinity distance directions.

Descartes also proposed an explanation of gravity whereby smaller particles accelerating to higher speeds tend to escape from slower, more massive particles, as to infer gravitational vacuum effect. However, it is contrary to the law of conservation of momentum that requires that whatever momentum comes in is the same that goes out. Even if higher speeds are created from within, a recycling effect of equal momenta is needed to maintain the more general presence of mass-energy. How it can be possible is according to the principle of entropy as different equilibrium states of existence having only partially visible effect of the internal process of change.

3

NEWTONIAN MECHANICS

Although Isaac Newton (1642–1727) has been considered the founder of classical Newtonian mechanics, there is nearly no part of it, if any, that had not been considered by someone else. His first two laws of motion, with regard to momentum and force, are attributable to the works of Buridan and Galileo. John Wallis (1616–1703) stated the second law in 1603, and the third law of motion, with regard to mutual action and reaction, was similarity offered by Leonardo da Vinci (1457–1515) in pointing out that air and water are equally resistant to each other. Robert Hooke (1635–1703) claimed he had suggested to Newton the inverse-square-law of gravity. However, it is not uncommon for ideas to be independently thought of by different thinkers. How such ideas develop into theory is what theorists are more apt to be credited with.

When Newton became a new member of the Royal Society, he did consult with Hooke, an elder member using the telescope along with a microscope, vacuum pump and so forth. Hooke had argued all bodies of mass move in straight lines until deflected by some force, and that the attractive force of gravity is stronger at shorter distance from its mass of attraction. Newton, himself, did admit that Hooke suggested the inverse square law, but it was Newton who actually formulated it according to his own formulation of calculus.

Newton stands out nonetheless as outstanding for his contribution in advancing theory. It was a comprehensive formulation of ideas that resulted in the unification of Kepler's heliocentric scheme of the solar system from which relative motion and gravity equate as forces of nature. Even though it was to be modified by the relativity aspects of space and time, his assumptions of absolute space and absolute time along with conservation of mass were contribution to a more comprehensive formulation of theory previous to that of special and general relativity.

Laws of Motion

In addition to the concepts of absolute space and absolute time as a means of determining events according to standard units of measure, Newton believed the material content of the universe always stays the same amount. It does not change by being in relative motion, under the influence of gravity, or by any means whatsoever. From this conservation of mass and assuming space and time are absolutes, he postulated three laws of motion:

1. Law of inertia: objects in a non-accelerating state of relative motion or at rest remain as such until they are acted on by an external force, such as by either gravity or collision with other mass.
2. Force is the product of a mass m and its acceleration $\mathbf{a}$ per time t with regard to acceleration.
3. An equal and opposite reaction occurs with every action.

The law of inertia is expressed as the product of a mass m and its velocity v as momentum P. Hence, the first law is according to the equation $P = mv$. By this law, momentum remains unchanged until acted on by another mass or external force.

With regard to the second law, the amount of force F used to change momentum is the product of mass $\mathbf{m}$ and acceleration $\mathbf{a}$, according to the equation $F = \mathbf{ma}$. Acceleration itself is either a change in velocity, either of speed or direction, or of both, per time. In ratio to mass, it is possible that a quick enough fly exerts more force from changing speed than someone throwing a brick. However, since the brick has far more mass than the fly, it generally has more force than does the fly.

Since velocity includes both direction and speed, the change in velocity can either be a change in speed or a change in direction or a change in both direction and speed. A rocket moving in a circle, for instance, constantly accelerates by means of constant force. The rocket can also constantly increase in speed by more force increasing the rate of circular acceleration.

Galileo had previously established these first two laws. Newton added the third one: the law of mutual force. Hence, if a force acts on a mass, the mass reacts with an equal and opposite force for a change in momentum of the rocket caused by the rocket fuel to result in an equal amount of change in momentum of the rocket fuel in the opposite direction, as the rocket fuel would otherwise be inexhaustible.

From mutual action and reaction between masses is conservation of momentum. Conservation means staying the same, and conservation of momentum means total momentum of all mass in any particular direction never changes by the action of one mass on another. The action can either be a collision of two or more masses or a force such as gravity. If a greater mass collides with a lesser mass, conservation of total momentum of the action is maintained by the change in velocity of the greater mass in one direction being relatively small and the change in velocity of the lesser mass in the opposite direction being relatively large. The mutual changes in each of the momenta is according to the equation

$$M(\Delta v) = m(\Delta V)$$

Change in velocity v of the greater mass M is thus less than the smaller mass m.

Consider our moon orbiting Earth, as depicted above, as an example of how equal changes in momentum of masses result because of their gravitational influence on each other being mutual. In this case, a relatively slower moving Earth being more massive than the moon results in a change in direction of the moon being a greater change in velocity at a greater speed. The moon's orbit around

Earth is thus a relatively large circular path, while Earth orbits within a small circular path of the larger one. As to why the Earth's path does not circle the moon, it is because Earth is too slow changing its direction at each new position of the moon.

Centripetal Acceleration

Centripetal acceleration is constant change in direction resulting in a circular path. Newton formulated it mathematically, but he delayed the publication of his work until 1974. Christian Huygens formulated it independently for his publication of it in 1673, but it was Newton who used it for the unification of celestial and terrestrial mechanics.

Consider, as by Newton's first two laws of motion, how a particle tends to move in a straight path, but it actually moves peripherally instead around a fixed point because of a centripetal force constantly acting on the particle in changing its direction. The centripetal force is any agent, such as gravity or whatever, preventing the particle from escaping its orbit.

Centripetal acceleration is a particular acceleration equating to any acceleration generally expressed by the equation

$$a = \frac{\Delta v}{\Delta t}$$

As to how it interrelates is explainable according to the equation

$$\frac{v^2}{r} = \frac{vr}{rt} = \frac{v}{t}$$

Accordingly, magnitudes v and t of v/t relate to centripetal acceleration as changes in direction per distance according to v and t. However, in comparison to different values of v and t, such as twice v occurring in one-half t around the same radius r, the result is that from moving twice as fast during half the time. A constant rate of change in direction around a circle of the same radius r, at twice velocity v in half time t, constitutes four times as much change per distance. Moreover, since the amount of speed v is inversely proportional to time per distance, it is speed squared per distance that determines the magnitude of centripetal acceleration. In other words, it is simply according to the magnitude of change in direction per time. The magnitude thus differs according to both speed and change in direction per distance. It is simply change in direction per orbital distance and time. Furthermore, the amount of change per time depends on the size of the circle. At a radius of one-half r, the amount of change is twice as much.

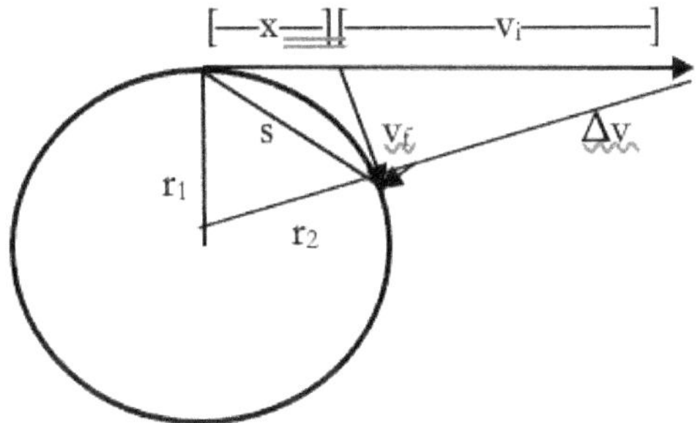

A geometrical interpretation is according to the above illustration. As illustrated, a system moves along an arc at constant speed v during time T from initial velocity v_i to a final velocity v_f, in relation to triangles. Radius r_1, an extended tangent from it to the right, and a radius r_2, extended to the tangent of r_1, form a right triangle. Another right triangle forms from the extension of the tangent of r_2 to the tangent of r_1. Because of the common vertex at the right, it is a smaller upside down and similar-right-triangle. All ratios of the corresponding parts thus equate. Because distance results from duration of speed, the vector directions v_i and v_f are representative of both distance and velocity. The other leg of triangle with legs v_i and v_f is a vector direction pointing towards the center of the circle to represent a change in velocity, and it is thus denoted as Δv. The ratio of $r_2 + \Delta v$ and $x + v_i$ of the larger right-triangle is the same as the smaller one, Δv and v_f, such that

$$\frac{x+v_i}{r_1} = \frac{\Delta v}{v_f}$$

The values x, v_i and v_f are interpreted as either speed or distance for $x + v_i$ and v_f to relate respectively as v and vT, where T is the time of acceleration.

These relations are according to a smaller distance during a less time of acceleration. For the smallest possible angle between radii, the arc between r_1 and r_2 converges with line segments s and v_f at the limit for shortest time of acceleration to equate in the manner

$$\frac{\Delta v}{vT} = \frac{v}{r}$$

$$a = \frac{\Delta v}{T} = \frac{v^2}{r}$$

Centripetal acceleration thus equals a constant change in direction towards the center of the circle.

Inverse Square Law from Kepler's Celestial Scheme

Since force is defined as F = ma, centripetal force similarly equates as

$$F_c = \frac{mv^2}{r}$$

Note: F_c increases for smaller r, as for faster change in direction at the same speed around a smaller circle.

Another force is gravitational, which is also partly centripetal with regard to it maintaining orbital motion. This connection was the means that enabled Newton to unify Kepler's celestial scheme with the forces of nature for formulating a theory of gravity.

Newton derived his formula of gravity by means of calculus, as he was one of its founders, but the formula can also be derived in a manner consistent with Kepler's planetary laws of motion.

Kepler's third law, in particular, relates the planetary orbits in our solar system. Accordingly, the period squared of any elliptical orbit equals the cube of the mean distance of the planet from the sun. The ratio of the time-squared of Earth's revolution around the sun to a cube of its distance from the sun (as from Earth's center of mass to the sun's center of mass) is thus the same as that of any other planet in the solar system.

Kepler's third law in mathematical terms is

$$A^2 = kr^3$$

The letter k represents a constant of proportionality for the proportionality between the period of revolution squared and mean radius r-cubed of the semi major axis of an elliptical orbit.

If the ellipse is a circle, which its average equates to, then r is representative of a circle. As for simplicity, let the distance of orbit be that along the circumference of a circle such that the orbital distance is the number π (pi) multiplied by twice the radius r of the circle. (An average length of the two foci of an ellipse equates as a radius of a circle, such that a circle is truly representative of the average distance of an ellipse, whereby the two foci have converged to the same position.)

The time or period of revolution of the planet can also be expressed in terms of distance divided by orbital speed v in the manner

$$A = \frac{2\pi r}{v}$$

Squaring and combining equations results as

$$A^2 = \frac{4\pi^2 r^2}{v^2} = kr^3$$

Multiplying the last two sides of the previous equation by v^4, and by dividing them by k and r^4, results as

$$\frac{v^2}{r} = \frac{4\pi^2}{kr^3}$$

Centripetal acceleration according to Kepler's scheme is thus proportional to $4\pi^2/kr^3$.

Multiplying both sides by m relates to centripetal force in the manner

$$F_c = \frac{mv^2}{r} = \frac{4\pi^2 m}{kr}$$

Since r is a common distance separating the combined forces of masses m_1 and m_2, they combine by multiplication for the total force to equate in the manner

$$F_g = \frac{4\pi^2 m_1}{kr} \cdot \frac{4\pi^2 m_2}{kr} = \frac{16\pi^4 m_1 m_2}{k^2 r^2} = \frac{G m_1 m_2}{r^2}$$

This equation expresses Newton's general form of the inverse-square-law of gravity, with G as a constant of proportionality in place of $16\pi^4/k^2$. Its determined value is $6.67428(27) \times 10^{-8}$ cubic centimeters per grams and per seconds squared.

Explaining Gravity

Newton was not content with his inverse square law of gravity only explaining action of one mass on another as occurring at a distance. He considered it as incomplete, and he attempted to explain it more completely in manner of a contiguous action of masses affecting the space between them for action between masses to somehow result in their mutual attractions.

Although Newton attempted to explain gravity according to an agent acting between mass for contiguous action to occur, he was still reluctant to recognize an ether filled space as the medium for wave action. He regarded its presence as an obstacle to the free movement of planets and the natural motion of mass in general unless it could be rare in content.

Newton probably would have reconsidered the ether if he had known that photons of light move through a so-called vacuum space at the same speed. How is that possible without some kind of equilibrium condition of a medium? He did consider extremely rapid moving particles of minute mass can exist for internal elasticity of containment. In theory, a particle of less mass can have a greater speed while it moves with the same momentum as does a more massive but slower-moving particle. An internal containment of the smaller but faster moving particles equally acting in all directions against larger but slower moving particles is in compliance with conservation of momentum. They can be maintained as equilibrium states according to containment by reflection of other particles. Inner particles can also be divided into even smaller particles allowed to escape. However, for gravitational effect, there then needs to be a reverse process,

The smaller particles that escape could be in an equilibrium state that is less detectable by a greater amount of them interacting with themselves. The reverse process could be a slighter tendency of them to unite by inelastic collision with larger particles. It is thus possible for there to be a greater number of faster-smaller particles outward pushing slower-larger particles inward.

Descartes had proposed an explanation of gravity whereby smaller, less massive particles would be accelerated to faster speeds for them to escape from the presence of more massive particles. However, Newton's laws of motion, according to his attempt to explain gravity according to direct action between particles instead of action at a distance, does not allow for a collision of one mass on another to result in one being pulled backward.

Another possibility to consider is that of wave effect instead of particle effect are part of the process. Whereas particles of matter cannot occupy the same space, waves superimpose for invisible effect of passing through an equilibrium state of existence. The existence of the equilibrium state, as inclusive of this possibility, is explainable according to the principles of entropy and covariance. The latter is that of relativity theory. The former is a particular form of energy conservation according to the second law of thermodynamics.

A Galilean Interpretation

The relativistic modification of Newton's inverse square law still maintains it as an essential part of theory. Another part of it is an explanation of gravity according to Galileo's principle all bodies of mass fall through the vacuum of space towards Earth at the same rate. It indicates gravity for each part of mass is proportional to the amount of mass gravitating. However, the numerical calculation depends on a chosen unit of mass, as illustrated below of two masses each divided into three equal parts.

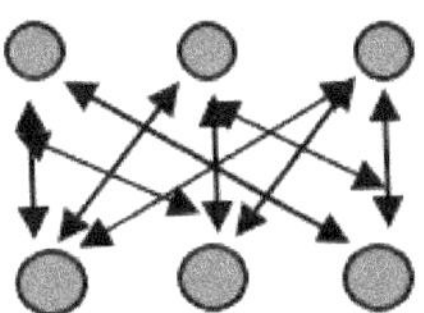

The total interaction of gravity between masses is thus simply a number of parts of one mass multiplied by a number of parts of another mass of equal magnitude. It relates to Newton's inverse-square-law of gravity in the manner

$$\frac{GM_1m_2}{r^2} = \frac{M_1v_1^2}{r} = \frac{m_2v_2^2}{r}$$

$$\frac{G(3M_1)(3m_2)}{r^2} = \frac{(3M_1)(3v_1^2)}{r} = \frac{(3m_2)(3v_2^2)}{r}$$

Three mass parts multiplied by three other mass parts thus equal nine centripetal forces even though their speeds of centripetal acceleration only increase by the square root of 3.

Note: the increased speed is the same for any two triple-mass quantities. What does change, for it to maintain the same amount of actual energy, is the numerical value of the gravitational constant G because of the difference in chosen units of measure. The numerical value of the gravitational constant G needs to be three times less for only a change in the numerical value of mass. Similar conditions apply regarding chosen units of time and distance.

The same condition applies to the measure of electric charge, but it can differ in actual values of internal parts because of positive and negative charges combining to neutralize each other. Quarks inside the nucleus of the atom, for instance, have one-third as much charge as is detected of more general particles of nature.

It might seem possible in theory for gravitational force to both decrease and increase in value for it to be able to cancel out if all mass became evenly distributed throughout the universe. However, there is still the internal force of gravity holding individual particles together along with electromagnetic force. A decrease in radius of the particles results in a reciprocal increase in distance between the centers of particles. As long as average density of mass stays the same, conservation of gravitational force applies. However, being in such an equilibrium state of order is contrary to conservation of entropy applying to the detection of disorder whereby the thermal radiation of two objects of the same temperature cannot directly change the amount of heat of the other.

Still, explaining the cause of gravity has significant relevance inasmuch as gravity has been excluded from standard theory uniting such other forces of nature as kinetic and electrostatic energies. Explaining how gravity complies with the laws of kinetic energy is thus an essential means of determining its unification with other theory.

Kinematics

An escape speed from gravitational mass M as part of Newton's inverse square law relates in the manner

$$\sqrt{\frac{2GM}{r}} = v\sqrt{2}$$

The escape speed is thus the square root of 2 more than the orbital speed.

This relation is explainable in terms of kinetic energy. The word kinetic derives from the Greek word *kinesis* that means motion. Although its concept could have been aware of, it seems not to have been applied until a much later time, as about 1850 by Lord Kelvin.

Gottfried Leibniz (1646–1716) called the energy of motion vis viva, meaning the living force. Willem's Jacob Gravesande (1688–1742) discovered the penetration of clay by equal weights that are dropped from different heights is in proportion to the difference in speeds squared of each weight. Emilie du Chatelet (1706–1749) explained this result as work energy of the change in speed of mass used to move a mass quantity the distance proportional to the inertial resistance causing the change in speed.

With calculus, Joseph Lagrange (1736–1813) found in 1811 that there is an additional difference of force mv^2 and a potential energy of momentum mv by a factor of 2. A factor ½ for kinetic energy $K = (½)mv^2$ came from Gustave Coriolis (1782–1843) in 1829, even though a kinetic theory of gases was developed earlier by Johann Bernoulli (1667–1748) in accordance with Newton's laws of motion.

Consider a movable partition separating two gases into equal volumes of cubic space. Each gas has the same number of molecules, but the molecules of one gas have only one-fourth the mass moving at twice the speed on the average than the more massive molecules of the other gas. Since $(1/4)m(2v)^2 = mv^2$, the gases have the same amount of potential energy, and since one-fourth mass at twice speed strikes the partition twice as often, the momentum of action on both sides of the partition per time of action is the same, such that the partition does not move.

Conservation of kinetic energy of elastic collision is proven with three equations according to Newton's laws of motion:

$$(1a) \qquad (nm)v_1 + mv_2 = (mn)v_3 + mv_4$$

$$(2a) \qquad \frac{1}{2}(nm)v_1^2 + \frac{1}{2}mv_2^2 = \frac{1}{2}(nm)v_3^2 + \frac{1}{2}mv_4^2$$

$$(3a) \qquad v_1 - v_2 = v_4 - v_3$$

Accordingly, n denotes any positive real number such that the product nm denotes any mass quantity as proportional to m. The left sides of equations (1a) and (2a) represent the momentum and kinetic energies of the masses before collision. The right sides of these equations represent the same after collision: velocities v_1 and v_2 are before elastic collision, and velocities v_3 and v_4 are after collision. Equation (3a) defines elastic collision such that the difference in relative speed between masses remains the same after collision as before collision, but they have been reversed in their original directions.

Equations (1a) and (2a) are divided by m, and both sides of equation (2a) are multiplied by 2 for simplification, obtaining

$$(1b) \qquad nv_1 + v_2 = nv_3 + v_4$$

$$(2b) \qquad nv_1^2 + v_2^2 = nv_3^2 + v_4^2$$

$$(3b) \qquad v_1 - v_2 = v_4 - v_3$$

To prove kinetic energy is conserved in elastic collision, the task is to derive equation (2a) from equations (1b) and (3b).

Equation (1b) rearranges by subtracting v_2 and nv_3 from both sides of it to obtain

$$nv_1 - nv_3 = v_4 - v_2$$

$$n(v_1 - v_3) = v_4 - v_2$$

By dividing both sides of the resulting equation by $(v_1 - v_3)$, a solution of n is obtained as

$$n = \frac{v_4 - v_2}{v_1 - v_3}$$

Rearranging equation (3b) by adding v_2 and v_3 to both sides of it obtains

$$v_1 + v_3 = v_4 + v_2$$

Dividing both sides of it by $(v_1 + v_3)$ obtains

$$1 = \frac{v_4 + v_2}{v_1 + v_3}$$

The product of the two solutions n and (1) gives

$$(1)n = \left[\frac{v_4 - v_2}{v_1 - v_3}\right] \cdot \left[\frac{v_4 + v_2}{v_1 + v_3}\right] = \frac{v_4^2 - v_2^2}{v_1^2 - v_3^2}$$

Multiplying by $(v_1{}^2 - v_3{}^2)$ obtains

$$n(v_1^2 - v_3^2) = v_4^2 - v_2^2$$

$$nv_1^2 - nv_3^2 = v_4^2 - v_2^2$$

Adding $nv_3{}^2 + v_2{}^2$ to both sides of this result obtain

$$nv_1^2 + v_2^2 = nv_3^2 + v_4^2$$

Multiplying both sides by m/2 obtains

$$\frac{1}{2}(nm)v_1^2 + \frac{1}{2}mv_2^2 = \frac{1}{2}(nm)v_3^2 + \frac{1}{2}mv_4^2$$

(2a) is thereby derived from (1a) and (3a) in proving conservation of kinetic energy in elastic collision from the conservation law of momentum and the difference in relative speeds after collision.

Conservation of kinetic energy is maintained by elastic collision in manner of also conserving relative motion. Twice mass changes the relative speed of the other mass twice as much by decreasing its relative speed half as much. The total increase in speed of one mass is thus the same as the total decrease in speed of the other mass.

Contrary to this analysis is inelastic collision whereby relative motion between mass appears to decrease. However, an inelastic collision is merely more complex, involving such internal motion related to heat and emission of such electromagnetic radiation as light.

There are also conditions of relativity theory to consider regarding the difference between mass and energy. According to special relativity, there is an exchange of mass between one mass as part of the exchange of momentum for less increase in speed. The exchange is thus partly inelastic whereby conservation of kinetic energy is even more complex. Moreover, there is an increase in mass per increase in speed such that a mass needs to increase to infinity to obtain light speed. In other words, it takes an infinite amount of mass to collide and be absorbed for all of it to reach light speed.

Mass is thus a particular form of energy that converts to or from another form of energy such as light. For conservation of mass to apply, another form of energy, if not light itself, needs to convert into mass. Newton did theorize mass converts to light and vice versa. It becomes relevant with regard to uniting gravity with electromagnetism. Massless light energy is assumed to influence electromagnetic energy. Gravitational energy is similar but also conditional to an equivalence principle of gravitational and inertial mass according to both Newtonian Mechanics and relativity theory, despite gravitons of gravity being massless.

The relativity between light and mass is also more complex regarding inelastic collision between mass and massless particles. For instance, according to special relativity, no conservation of momentum occurs of an elastic collision of a mass increasing in relative motion by completely reflecting a photon in the opposite direction. For instance, if mass m at rest obtains three-fifth light speed by absorbing a photon, it cannot, according to special relativity, obtain six-fifth light speed by completely reflecting the photon to move in the opposite direction. The massless photon is here illustratively used in compliance with a massive one.

It is not that a photon can be completely absorbed by mass; it is that the absorption is conditional to a Doppler effect whereby the amount of absorption and reflection of the photon is in proportion to the difference in velocities of the two interacting particles, whether the two particles are both material particles or one is that of light energy.

It is also noted that the amount of light absorption is also dependent on such other factors as color. For instance, if it is possible for a substance to exist as pure black, then it supposedly would absorb a photon in its entirety according to more complex conditions apart from relative motion. Similarly, pure white substance would reflect the photon in its entirety. However, that is assuming pure white and black colors can exist for black not to change by complete absorption of light, or for white to completely reflect all the light it encounters. It also considers the possibility of light having the property of momentum. How can light and gravity change momentum of mass if photons and gravitons are themselves massless?

4

KINEMATICS ATOMS
AND THERMODYNAMICS

Sometime in the thirteenth century Giles of Rome (b. before 1247–d. 1316) proposed an atomic theory according to a condition that no form of matter exists smaller than a minimal quantity of substance. He tried to support his theory with the investigation of a vacuum state. He was unable to verify his theory, but investigations of the vacuum state continued, eventually leading to Boyle's Law, the kinetic theory of gasses, development of chemistry, laws of thermodynamics, and the classical theory of the atom.

Boyle's Law

Giovanni Batiste Beliani (1582–1666) debated with Galileo on how to explain vacuum effects. In the year 1620, the debate focused on air weight. It was known water could flow higher up from a vessel through a tube lying over a hill. However, if the top of the vessel was sealed, partial vacuum occurred at the top of the vessel from leaked water at the bottom, as to restrict the water from flowing through the tube. It was suggested by Galileo there are attractive forces between the water and the vessel, whereas Beliani believed the cause was outside air pressure on the water attempting to come out of the tube at its other end. Beliani was correct, as further investigation of vacuum states led to such new inventions as the mercury barometer by Evangelista Torricelli (1600–1647) and the air pump by Otto von Gueriche (1602–1686).

Torricelli used the barometer to compare air pressure at sea level to its pressure higher up in the mountains. He not only found a difference; he further discovered the pressure changes with a change in the weather. With mercury thirteen times denser than air, it is able to create a vacuum in a tube that varies according to present pressure of the atmosphere.

Gueriche invented the air pump to produce more vacuum in order to measure more work capacity of outside air pressure. The outcome was him using the barometer in 1660 for forecasting the weather. He also published books that Boyle read, such as one defending the vacuum state of existence accepted by Galileo and Newton and opposed by Descartes.

More experimental results occurred in England with Henry Power (1623–1670) and Richard Townely (1629–1707) more generally verifying the volume-pressure-product constant occurring at various altitudes. Boyle then proposed, in 1662, the law that the product of volume V and pressure p of a gas being constant at a particular temperature is expressed by the equation $pV = k$. (Also, in France, this law was proposed in 1676 by Edme Mariotte (d. 1684).)

The constancy of pressure per volume relates to Newton's laws of motion in that a sphere with twice the radius of another having eight times more volume, is consistent with masses of twice radius having eight times more centripetal force of gravity between them: $Gm^2/(2r)^2 = (v^2/4)(2r)$. The intensity of gravitational attraction between two larger masses of one-eighth density and one-eighth centripetal force is thus consistent with Boyle's law.

What followed from Boyle's law is ways of relating heat and temperature. Although the motion of atoms or molecules causes them, what physicists essentially knew about them at the time was in relation to heat. A quantity contained of mass and temperature is a qualitative measure of how much a particular substance such as mercury expands in relation to heat absorbed, but Boyle's law indicated a way of understanding them in mechanical terms of relative motion and mass.

As was noted, constancy of pressure per volume is explainable according to kinematics. Daniel Bernoulli (1700–1782) initiated the kinetic theory of gases along with his study of hydrostatics. Bernoulli advocated a mechanical theory in analyzing the kinematics of molecules in view of particle collisions despite a general regard at the time that such a process is too simple to resolve the more complex nature of reality.

Although the pressure per volume constant was determined constant for a particular temperature, it was still questionable as to whether the constant k is the same for different temperatures. Guillaume Amontons (1663–1705) foresaw the ideal gas law pV = nkT before it became published in a 1702 paper stating the product of pressure p and volume V is the product of temperature T for the same constant k, such that either pressure, volume or both increase with the increase in temperature. He further considered the zero temperature in relation to zero pressure, which anticipated an absolute temperature scale established about a century and a half later.

Another form of the law, Charles' law, became a law of volumes whereby the volume of a gas container increases with temperature instead of an increase in pressure. The law was proposed by Jacques A. C. Charles (1746–1823) in 1787. It along with Archimedes' principle of buoyancy led to the invention of the hot-air balloon. It was also established quantitatively by Joseph Louis Guy-Lussac (1779–1850), in an 1802 publication, whereby one-degree centigrade change in temperature corresponds to a change in volume of the same pressure occurring as one part in 273 parts of the gas volume.

An ideal gas law was hinted at by Amontons in 1702. It was explicitly stated in 1834 by Paul Emile Claypeyon in relation to Boyle's and Charle's laws. August Karl Kronig (1822–1879) derived it in 1856 in accordance with the kinetic theory of gases, as Rudolf Clausius did as well in 1857. However, Johannes van der Waals (1837–1923) disclosed, in 1873, the law is not ideal because of internal effects influencing the results. For instance, the number of molecules can be influential by them reversing their directions from colliding with each other. Even with more complex formula, predictions of it are generally only reasonably accurate with gases of higher temperature and lower pressure.

A criticism of the kinetic theory of gases itself was that internal motion of matter would likely cause it to explode every which way. It would not be until the middle of the nineteenth century until a counter argument would come forth with Clausius explaining that collisions between minute particles great in number obstruct their mean free path of escape. Constant collisions keep reversing directions for the total distance moved by a particle to be longer than the direct outward distance itself. However, more reasonable is that mutual interaction of particles results in different equilibrium conditions of other forces. The kinetic theory of gases was thus not to be accepted until revived in a statistical form by Clausius, Maxwell, Boltzmann and others in a later part of the nineteenth century.

The explanation of Boyle's law itself is simple and conditional to the same amount of average speed per particle, and it is significant as to how it relates to another law, the Stephan-Boltzmann fourth power one that led to the formation of the Planck constant for further formulation of atomic theory.

The Substance of Heat

Heat as molecular motion had many proponents, including Boyle and Hooke, but it was not until the later part of the nineteenth century for it to become a more generally accepted theory. The criticism pertaining to an inner violent molecular motion causing matter to explode in all directions was influential. For this reason and others, the idea heat is a particular substance absorbed and emitted by matter instead of only being an internal movement of the internal components of matter remained popular. However, debate on the nature of mass and heat continued its advance.

Both Boyle and Newton proposed fire consists of material substance as the residue caused to burn and produce heat. Boyle experimented to find substances that did not decompose, which he defined to be an element. The concept of the element eventually led to the discovery of the atom.

Etienne Francois Geoffrey (1672–1731) advanced, in 1718, the idea that a particular substance of a compound (such as carbon of carbon monoxide) is replaceable by another substance (as by hydrogen to convert the oxygen of carbon monoxide into water molecules). He then contrived a table of sixteen columns in demonstrating an order of replacements of known substances. This effort evolved into a table of affinities whereby element A instead of element B combine with element C because of elements A and C having more of an attraction for each other than do elements B and C.

The table of affinities was utilized by Joseph Black (1728–1799), Henry Cavendish (1731–1812) and Joseph Priestley (1731–1810) to discover nearly all elements of permanent gases. Their discoveries led to an explanation of a weight oddity from combustion and calcinations of dissimilar materials that result from a combination of different elements further resulting in various combinations of exchange.

Newton had considered the existence of an inert matter that is contained by elastic forces of some kind of material medium. He also offered an explanation of heat as a repulsive force decreasing in inverse proportion to distance between molecules. He argued light particles excite the containment of inner particles, which then conveys the excitement as heat effect. He supported the notion of heat as the inner motion of particles needed to overcome vacuum effect, which could explain, in part, the attraction of gravity. However, the other part would likely be the presence of a positive force in otherwise vacuum space.

Attraction and Repulsion

Newton was also influential to future theory for advocating a dual principle of both attractive and repulsive forces of nature: one being gravity and the other as heat. Stephen Hales (1677–1761) developed an idea of both attractive and repulsive forces consisting of two kinds of matter tending to become balanced in a state of equilibrium. His idea was to have a profound influence on Benjamin Franklin (1706–1790) and others with regard to two fluid theories of electricity and magnetism.

Theorists were generally inclined to relate all forces as substances of a particular kind determining the internal natures of matter. In the 1740s, for instance, Franklin proposed that electrical phenomenon is an elastic fluid of mutually repulsive particles. Matter is electrically neutral for containing the right amount of a particle fluid, repulsive if it contains an excessive amount, and attractive if deficient of it. Gowin Knight (1703–1772) also proposed a fluid for magnetism with the propagation of light as the vibrant motion set up in the fluid.

William Cullen (1710–1790) advanced the idea natural forces result as various modifications of ether. He proposed electricity, light, heat, gravity, magnetism and so forth emanate as various ether forms that are themselves gravitational weightless and distinct from matter, thus allowing additional effects deviating from an equilibrium state of gravity. Such a proposal was in line with the beliefs of Gassendi and the public in general.

Latent Heat

Cullen's student Joseph Black (1728–1799) also considered heat, light and so forth as modifications of ether. In support of his consideration, he systematically studied combustion and calcination of different elements in pioneering the science of calorimetry.

Black's study pertained to temperature and heat. Newton had defined a quantity of heat as the amount of time taken to lower a substance to room temperature. Black took this definition to mean that heat can be measured as a time required for either dropping or raising its temperature to a certain degree. He measured a certain quantity of water according to the amount of time it takes to raise its temperature one degree. The temperature, however, did not change in such cases as involving change from water to ice, or vice versa, and from this find, Black proposed the concept of latent heat.

Latent heat had previously been noticed by Daniel Gabriel Fahrenheit (1686–1736). He discovered water remained in the liquid state while cooling at the freezing point (as thirty-two degrees Fahrenheit), but it congealed into a solid state above this temperature by means of shaking the container. This discovery that Black verified on his own in 1761 suggested heat can be stored as a latent form of energy without a rise in temperature. A mixture of ice and water, for instance, need not change in temperature with change in heat if a change in heat is relatively slight.

From Phlogiston to Caloric

In an effort to disprove a theory that Phlogiston is a primary substance of fire, Antoine-Laurent Lavoisier (1743–1794) probably introduced caloric in naming the main substance of heat. George Ernst Stahl (1660–1724) had proposed Phlogiston is a substance of violent motion producing flame and heat when entering into the air from the dispersing of matter. Plants absorb and then recycle it. There is partial truth to this idea if we identify Phlogiston as carbon, which Stahl did, but he also claimed it is the only element existing not as a compound but as a primary substance and a catalyst for all processes of combustion.

Stahl's generalization was open to dispute since some substances lose weight during calcinations whereas others gain weight. Thus, in some cases, Phlogiston needs to be of positive weight; in other cases, it needs to be of negative weight.

A resolution to this quandary was the table of affinities first presented by Geoffrey wherefore came many chemical discoveries. One of particular pertinence was the discovery of oxygen. Because of it, physicists no longer needed Phlogiston to explain changed states of matter. Calx-plus-Phlogiston-producing metal, for instance, could more easily be explained as calx giving up oxygen to become metal. Conversely, metal absorbing oxygen becomes calx. Furthermore, experiments by Lavoisier indicated conversion of sulfur, phosphorus, and arsenic into oxides results in a gain in their weight and a decrease in the weight of air.

In disclosing the inadequacy of Phlogiston theory, Lavoisier held onto the general belief of Cullen and others that ether acts as a weightless elastic fluid that is not influenced by gravity and that it is primary responsible for such effects as light, heat, electrostatic repulsion and so on. He followed up on experimental results of Cullen on cooling effects of such volatile liquids as alcohol by vaporization to discover, independently of Black and Fahrenheit, that temperature of ice did not increase while changing into the liquid state. Along with replacing Phlogiston with caloric, he proposed an element such as oxygen has more affinity for absorbing caloric than does another, which could also allow ice to change to water while maintaining a slight portion of its coolness by means of a base material. The caloric thus seemed to provide more consistent explanation of how change occurs of substance, but it was to be overcome with the development of atomic theory and the laws of thermodynamics.

From Caloric to Atomic Theory

It was evident substances combine in definite proportions according to a precise amount of oxygen combining with a precise amount of hydrogen. This find was named by Joseph Proust (1754–1826) the law of definite proportions. It and a similar law led the way to modern atomic theory.

The similar law is of multiple proportions proposed by John Dalton (1746–1844) whereby chemical elements consist of tiny particles called atoms. Atoms of a particular element are all of the same size, weight, mass, etc. that differ from those of other elements, but the different elements combine in ratio of whole numbers to form chemical compounds. Dalton proposed a numerical list of atomic weights of six known elements: hydrogen, oxygen, nitrogen, carbon, sulfur and phosphorous. Hydrogen, as the lightest of these elements, was assigned the number one.

Another similar law, proposed by Joseph Louis Gay-Lussac (1778–1850), is the law of combining proportions. He and Alexander von Humboldt (1769–1859) discovered that two volumes of hydrogen combine with one volume of oxygen to become two volumes of vaporized water having the same temperature and pressure. Gay-Lussac further studied data collected by Humphrey Davy (1778–1829) with regard to volume ratios obtained by combining nitrogen with oxygen. He experimented to further discover a half gaseous volume of nitrous oxide is obtained by combining a volume of nitrogen and a half volume of oxygen of the same temperature and pressure.

Similar finds encouraged Gay–Lussac to conclude that gases combine in whole-numbers ratios as they also do according to weight. However, he did not explain why some whole-number ratios differ from others. Why, for instance, does water vapor squeeze one volume of oxygen and two volumes of hydrogen into two volumes total?

Amedeo Avogadro (1776–1856) explained the ratios of whole number combos according to a hypothesis gas with the same volume, pressure, weight and temperature contain the same number of molecules according to a numerical constant. As to why two volumes of water vapor is the result of two volumes of hydrogen gas and one volume of oxygen gas, he assumed molecules are formed from "solitary elementary molecules", which became known as atoms. Since hydrogen gas contains two atoms for every molecule, two hydrogen atoms combine with an oxygen molecule as a single atom to form into water consisting of two different molecules. The number of molecules per volume thus stays the same in allowing the number of atoms to vary instead.

Avogadro's hypothesis became his law proposed in 1811, but it was at odds with Dalton's atomic theory that assumed the compounds of elements result from like atoms repelling each other in not allowing them to occupy together the same space. Contradictions to the law also appeared to occur at certain temperatures. Thus, Dalton and his followers were not to accept the concept that identical atoms combine in becoming a molecule. As it were, Avogadro's law failed acceptance until a fellow Italian, Stanislao Cannizzaro (1826–1920), pointed out in 1861 that the law could be used for a convenient table of atoms in simple ratios of whole numbers, and the apparent contradictions became explained and verified by experiment according to particular chemical reactions with temperatures in particular.

A particular significance of the Avogadro constant is the number of atoms or molecules of a gas per volume is the same for any particular temperature and gravitational weight. It provided a mathematical means for their determination of their mass, size and so forth.

Rudolf Clausius (1822–1888) later helped promote the kinetic theory of gasses in explaining the vibrant motion of atoms is stable by being great in number, whereby each encounter between numerous atoms acts to slow the mean free path of escape by means of them slowing down on impact to reverse directions, which is consistent with an interpretation of gravity by Newton according to vibrant cells having zero total momentum internally for each cell but which implies that space is indeed filled with an enormous number of miniature cells as a medium of interaction.

The Fate of the Caloric

The role of caloric in developing theory was to replace the phlogiston of fire as an explanation of latent heat. Properties ascribed to caloric where it consists of weightless particles repelling each other to flow from hot to cold matter. Because caloric was also assumed to be conserved, an engineer named Nicolas Leonard Said Carnot (1796–1832) used the caloric theory for the derivation of theorems to explain how a steam engine is able to operate more efficiently.

Postulating conservation of heat in view of caloric, Carnot theorized it flows from hottest to coldest parts of the mechanism performing work, caused by the viscous flow of caloric. Since caloric is conserved, the process compares to a waterwheel turning by the flow of water that forever recycles. Similarly, because systems of the same temperature are unable to exchange caloric to perform work, recycling of caloric is required. Adding more fuel, for instance, allows a hot steam engine to continually release caloric for it to perform work.

Carnot's condition for efficiency was correct for the most part, but the conservation aspect of caloric is inconsistent with conservation of energy in that energy merely converts from one form to another. Caloric was thus to become discarded in favor of energy conservation.

Neither physicists nor chemists had yet established the modern law of conservation of energy, but Count Rumford (1753–1814) had demonstrated an enormous quantity of heat results from boring cannon holes. Humphrey Davy (1778–1829) rubbed ice plates against each other to demonstrate heat is producible below freezing temperature even though no caloric should be available from it. Moreover, no appreciable amount of change in mass or of its weight occurred in either of these experiments. Rumford thus proposed vibrant motion causes heat instead, whereas Davy considered heat occurs from absorption of light. He also suggested that light combines with oxygen to become *phosoxygen*, a process whereby mass increases by absorbing light.

Conserving Energy

The main difference between conservation of caloric and conservation of energy is caloric does not convert into other forms of energy whereas all energy does, according to its modern concept. This modern concept was stated in 1841 by Julius Robert Mayer (1814–1878) as a force (then regarded as a varying form of energy) that merely changes from one form to another. It is thus neither created nor destroyed. He argued the loss of kinetic energy during inelastic collision between masses transforms into heat as an internal form of continuous motion.

James Prescott Joule (1818–1889) verified Mayer's argument in relation to friction. Heat that results from stopping motion by friction was common knowledge. Joule measured it quantitatively in relation to the magnitudes of heat from work used to overcome friction. It was also known that the flow of electricity through a wire tends to heats the wire. Joule established this effect quantitatively in 1841 in confirming electric current transforms into heat in mechanical units of work.

In 1847, Hermann von Helmholtz (1821–1894) addressed the scientific community in stating there is no such thing as a perpetual motion machine performing work without compensation for it in return, except for the universe as a whole. Explanation refers to an isolated system in that the internal energy of the system comprises, as stated in modern terms, a total amount of kinetic and potential energies of the molecules remaining constant until acted on by some external influence. Whenever interaction occurs, a change in internal energy is according to the quantity of heat absorbed as the amount of work performed on the system according to the equation

$$\Delta E = H + W$$

H denotes quantity of heat, W denotes the amount of work performed, and E proceeded by the Greek delta letter Δ denotes change in energy.

Note that one form of energy relates in terms of its potential. A ball at some height, for instance, has a gravitational potential as weight that can be converted into kinetic energy. If it falls from the table, then kinetic energy further encounters friction of the ball falling through air and colliding with the floor, converting further into heat energy. Potential energy thus links to force regarding storage of energy. Moreover, there is also entropy as useful energy to consider that can lose its usefulness unless compensated by some change outside its acquired state of equilibrium.

Entropy

Carnot's theorem became reconsidered according to a concept of entropy as the second law of thermodynamics. The law was proposed in 1850 by Emanuel Clausius (1822–1888) who mathematically formulated it in 1865.

Entropy became referred to as a measure of disorder required for change. William Thompson (1824–1907), who was renamed Lord Kelvin, had provided an example of it, in 1951, in terms of thermodynamic energy.

In view of energy conservation in contrast to the caloric, adding fuel to a system is to sustain a difference in temperature for doing work. However, there are various forms of energy, and there is a distinction between useful and non-useful energy. Useful energy produces change; non-useful energy is in an isolated state of equilibrium not changing unless by outside influence. Consider, for instance, two bricks of the same temperature being in thermal equilibrium. They are unable to initiate thermal work on the other by means of exchanging heat from one to the other, but if one brick encounters a cooler climate, then exchange of heat between them is able to occur. Disorder as difference in temperatures thus allows change to occur.

It is therefore possible to have a certain amount of energy in the form of heat at absolute temperature, say T_1. In theory, we can only harness all of the energy of a system if its temperature is reduced to absolute zero. Generally, however, systems are somewhere between absolute zero whereby the energy can only be harnessed by lowering it from T_1 to T_2. Such an example of entropy was explained in 1851 by Lord Kelvin to occur in the manner

$$1 - \frac{T_2}{T_1}$$

Applicable to this condition is an absolute temperature scale, which Kelvin introduced after Joule suggested in a letter to Kelvin that it was possible to measure the difference from absolute zero. All heat energy is thus available if $T_2 = 0$; none is available if $T_2 = T_1$.

Since temperatures tend to even out, some theorists speculated to the contrary of the second law of thermodynamics that the universe could be trending towards a heat death of the same average temperature. However, such speculation needs to be consistent with kinematics and gravity. Although an inelastic collision can occur between masses for cancellation of their difference in speed, internal energy is created in the form of heat or some other kind of internal energy. In terms of the difference in gravity and light speed, for instance, is the special relativity condition of the relativistic factor squared $(1 - 2Gm/rc^2)$. If gravitational escape speed squared, $2Gm/r$, equals light speed squared, c^2, then they cancel each other out.

Another example for consideration, contrary to entropy, is that it could cancel itself out with an even distribution of mass throughout an infinite universe. However, if various systems of equilibrium maintain instead, gravity then only cancels out from a particular mass's center while its gravitational influence still occurs towards its center and between other masses at different locations.

Other speculation was that of entropy being conserved. If emitted radiation is gradually recycled back, for instance, then entropy is conserved as an adiabatic process according to definition. Such conservation of entropy was used by Ludwig Boltzmann (1844–1916) to derive a fourth power law that led to the Planck constant of quantum physics.

According to the mathematical formulation of entropy by Clausius, heat quantity Q in ratio to absolute temperature T is a determining factor of entropy of internal bodies of an isolated system whereby different temperatures determine a quantity of available heat for useful work. Consider an exchange of heat from two different bodies of heat at different temperatures whereby the resulting entropy ΔS from the exchange becomes

$$\Delta S = \frac{Q_2}{T_2} + \frac{Q_1}{T_1} \geq 0$$

If the system performs work, the amount of heat lost will generally differ from the amount of heat gained, but if the system is able to maintain in an equilibrium state of containment, then $\Delta S = 0$ for conservation of entropy as an adiabatic process.

Order and disorder are both attributes of reality. An equilibrium state of order could even be dependent on the amount of disorder apart from it. The restoration of an equilibrium state of an atom, for instance, could be more probable due to a greater number of options of its surrounding environment.

Entropy thus constitutes a complex method of analyzing the application of energy. Hawking, for instance, challenged the singularity condition of the black hole according to general relativity. There could be a greater probability of escape with the inclusion of Hawking radiation as part of the greater amount of disorder contained within the gravitational field.

Although black holes are now evident according to astronomical observation, there is also the paradox of the observable part of the universe expanding from once being contained within a tiny volume of space. Theory thus seems incompletely explained whether true or not.

Overall significance of entropy pertains to the existence of equilibrium states. Constant light speed is part of one. The quantum state of the atom is another. Electrostatic and gravitational forces could also be of particular equilibrium states.

Beyond direct observation are virtual particles for possible explanation of various effects as to how these different equilibrium states interrelate.

Similar to gravity is electric charge. It was formulated by Coulomb as an inverse square law, but it differs from gravity in that it varies in its casual effect, positive against positive and negative against negative charges being repulsive, and positive against negative charges being attractive.

Negative and positive charges can accumulate apart from each other, but there is also a quantum condition whereby parameters of both the electromagnetic constant of charge e squared and the Planck constant h relate to a particular speed v similar to but less than light speed c. With other parameters being of mass m and radius r, an increase in m is countered by a decrease in r such that the density of m increases to the fourth power. It is an equilibrium condition different from that of gravity and relative motion in that change occurs in discrete units instead of gradually. As with the latter, it is also conditional to relativity theory according to covariance whereby an observer moving at any speed relative to another can be relatively at rest whereby light speed is determined to be the same no matter what the relative speed of the observer is.

Democritus (461–370 BC) proposed, back in the fifth century BC, that all of reality, including light, is comprised of individual sub-components. Quantum mechanics now includes virtual particles that are not directly observable but are necessary to explain particular atomic effects according to either wave or particle action. Discoveries of certain effects have not been explained by either of these two actions alone. The wave-particle duality itself has become instead an essential part of general physics theory in analogy to waves of the ocean occurring from a liquid state of material particles. Equilibrium states of virtual particles conditional to entropy are thus typical of the laws of nature. They require recycling effects from matter partially converting to virtual particles and other virtual particles converting to matter.

5

FROM WAVE THEORY TO RELATIVITY

Theories of nature evolved from opposing views pertaining to how space is filled. By one, the internal nature of matter comprises indestructible atoms moving by way of a partial vacuum. By another, a primary substance fills all space as a plenum.

A belief in the plenum was popular from the time of Aristotle until the development of fundamental laws of motion. After the establishment of the Copernicus heliocentric view of the solar system and Galileo and Newton establishing their mechanics, Earth moving freely as though space before it is empty except for it only being partially filled with tiny indestructible atoms became more acceptable. Nonetheless, such philosophers as Descartes proposed, to the contrary, atoms are tiny vortexes of circling ether to merely appear as though they move as particles through empty space.

Nowadays many physicists consider the ether as nonexistent or invalid since it is invisible in theory and since physics confines itself to what has more observable effect in calculable manner, but the ether has nonetheless had a long history in the development of theory from which wave action is an integral part of, as illustrated by the chain reaction below.

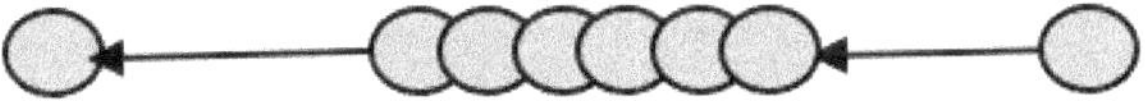

The illustration depicts how momentum continues either by a medium or empty space before it. Momentum through the medium is transferred to and from each iron ball by means of impulse. There is thus a momentum of impulse moving from one end of the row to the other when the former is struck by another ball. The wave momentum continues until reemerging as relative motion of another ball particle.

Particle action is evident, but wave action is also helpful for a more insightful understanding of the complexity of nature. Newton, for instance, was unable to explain the cause of gravity by way of contiguous action, such as with regard to how a particle striking another one can cause it to contract instead of being impelled in the opposite direction. Although it can be explained according to an "action at a distance" principle, he disallowed it in favor of empty space not being disruptive to planetary motion. Although other attempts have been offered to explain the direct cause of gravity, none have been accepted as decisive enough for Standard Theory. Wave theory can nonetheless be more extensive in its approach. Waves, for instance, can superimpose to negate their repulsive effects of direct action. Gravity can thus be explained as acting at a distance according to the propagation of gravitational waves through an otherwise undetectable medium.

Wave action, however, is extremely complex with regard to the many ways it can occur, depending on the nature of the wave-producing medium. Sound waves, for instance, propagate longitudinal action, whereas transverse action is more typical of the wave property of light. Most waves are periodic in nature, such that periodic motion of a particle in general can be described in accordance with a wave equation, but surface waves of the ocean vary in their periodicity with regard to a change in water depth.

Wave action can be insightful for anyone able to understand its complexity, but much of its complexity is here avoided in favor of restricting it to a history of light. Although the ethereal medium for light propagation is controversial in being itself directly unobservant, it is explainable according to the principles of entropy and covariance according to its disorder being an equilibrium state appearing the same to observers in different inertial states of relative motion.

Light and Wave Theory

Waves are noticeable events, such as surface ripples on a pond that are created by some sort of disturbance. In relating their wave action, Aristotle proposed light occurs from a wave-like disturbance of air. However, as far as is known, there was no constructive wave theory of light until 1678 when Christian Huygens (1629–1695) proposed his theory.

Huygens offered a principle of light waves vibrating perpendicular to their direction of motion at every point of space they contact as being itself the source of their spherical formation. He explained their properties of reflection and refraction, but another one of their properties he did not explain was that of diffraction.

Diffraction as a property of light was named in 1660 by Francesco Maria Grimaldi (1618–1663). He discovered the spreading of light passing through a small opening or slit results in a greater amount of spread of light moving forward. It supported the wave theory of light.

The law of refraction was accurately explained in a manuscript by Ibn Sahl (940–1040) of Bagdad as early as 984. In 1021, it became promoted in a treatise on optics by Alhazen (965–1014), who viewed light as consisting of rays of particles. It was first formulated in Europe by Willebrord Snellius (1560–1626) to become known as Snell's law.

Descartes later introduced the sine function for describing the ratio of angles according to refraction. Accordance to it, the direction of a stick submerging into water appears to change, but the change is an optical illusion. Instead of the stick changing direction, the light from it has changed directions twice by entering into and leaving from a denser medium, as refracted. Whereas reflection is simpler in that it equals the angle of incidence, refraction entails change in speed of waves entering into a different medium from which the ratio of angles of incidence Θ_1, and refraction Θ_2, equate with the ratio of speeds v_1 and v_2, and with the inverse ratio of refractive indexes n_1 and n_2 of the two mediums:

$$\frac{sin\theta_1}{sin\theta_2} = \frac{v_1}{v_2} = \frac{n_2}{n_1}$$

Huygens used his wave theory to explain a double refraction of light that was discovered by Rasmus Bartholin (1625–1698) in 1669 to occur in the calcite crystal called Iceland spar. In double refraction, the light rays split into two directions, one ordinary and the other extraordinary. The ordinary direction complies with the law of refraction whereas the extraordinary direction is a non-compliance of the law. An explanation of this noncompliance is given by Huygens' theory as ripples of the medium spreading waves in all directions. Even more complete explanation was then given with the development of wave theory to include properties of interference and transverse vibrations.

The corpuscular theory of light remained in favor until Thomas Young (1773–1829) proposed a principle of interference. Leonardo da Vinci had observed water waves are able to cross paths without obstructing each other's movement. Young applied this wave effect to light waves, and he included such additional interference effects of waves moving through other waves. He surmised waves superimposing combine to either increase or decrease their effect depending on the degree they are either in or out of phase. If waves overlap in phase, then they combine their effect. If waves overlap in opposite phase, then their effect is canceled for a dark spot to appear.

Young explained light and dark fringes in diffraction patterns of waves according to his principle of interference. Francesco Maria Grimaldi (1618–1663) had discovered sunlight spreads abnormally when it passes through a small hole. Young drilled two holes for an experiment from which he found light and dark fringes occurred when they otherwise would not if only one hole was drilled. The fringe pattern is explainable as the diffraction patterns of two rays of light overlapping in and out of phase. Where they overlap in phase, light amplitudes combine effect; if they overlap in opposite phase, no light appears. (The amplitude in this classical sense refers to the height of the wave from crest to trough.)

Some experiments produce no interference, which is now explained as light emitted from the source as relatively incoherent. Since light waves are extremely rapid and since trains of them are relatively short, it is possible to detect two or more rays superimposing only in particular circumstances. It occurs in Young's experiment by drilling holes close together. Another method is to split a light ray into two components for partial reflection. The split parts move through different paths of slightly different distances for them not to rejoin in their original state. They then superimpose instead into a slightly different state than their original one.

Transverse Waves

Young was able to explain nearly all light effects except polarization, a condition of opposites whereby particular direction—up-and-down or sideways-back-and-forth—is perpendicular to the direction that the waves move forward. To the contrary, longitudinal sound waves are density compression that move forward with no noticeable transverse effect. In 1808, Etienne Louis Malus (1775–1812) discovered that either reflection or refraction can produce polarization. Young had not explained it because his medium for light waves compared to the propagation of sound through air as longitudinal waves of rarefaction and condensation of the air medium.

Young did suggest in 1817 that waves needed to contain transverse components in order to explain polarization effects. Augustin Fresnel (1788–1827) formulated his theory of optics in 1818, which did explain polarization as transverse waves. His theory includes Young's principle of interference and also the principle set forth by Huygens of continuous waves spreading from all the points of contact with space. However, the transverse property of waves led to another enigma inasmuch as it does not normally occur in a three-dimensional solid state. A rigid medium for three-dimensional space seemed to be too much of an obstacle to explain how planets and other objects move as freely through it as they appear to do.

Possible explanation is according to entropy of an equilibrium state of underlying existence. However, entropy was not yet understood as such. By it, more complex interaction occurs within the equilibrium state of the medium for a perpendicular change in direction to occur in manner consistent with conservation of momentum. The transverse action just needs to reverse back and forth between two different equilibrium states.

Although the law of entropy had not yet been established, explanations similar to it were being put forth for casual explanations that eventually led to Maxwell's theory of electromagnetism.

The Elastic Medium

In investigating properties of the elastic medium, Claude Louis Navier (1785–1836) assumed objects are of extremely minute and compact particles whereby attractive and repulsive forces maintain in a state of equilibrium. The restoring force is analogous to a liquid reacting according to the motion of the particles. The solid state is conditional to the distance of separation between particles.

Navier's theory was elaborated on by Augustive-Louis Cauchy (1789–1857) with a law of elasticity that Robert Hooke had proposed. According to it, the stretching of an elastic body is proportional to the force applied for the stretching.

Cauchy interpreted Navier's theory as setting up a condition of strain. Physicists name the mathematical formulation of this condition a tensor in relation to a more complex vector quantity in applying to variable forces of higher order. A vector refers to a quantity having both particular magnitude and direction. A current, for instance, displaces the path of a boat crossing a river at a given speed. The vector thus pertains to both the boat moving in one direction and the current moving in another direction. Further tensor application could include an increase in the speed of the current, say, due to it nearing a waterfall.

Cauchy's results were mathematically consistent with those of Navier's homogeneous media, but more than one elastic constant of proportionality is needed for isotropic media. Whereas a homogeneous medium is the same everywhere, an isotropic one is the same only in directions that can vary in distance and other aspects. The media is necessarily isotropic in the case of Cauchy's results in allowing more than one kind of wave (transverse and longitudinal) to propagate through it.

The equations of elastic solids were incompatible with optics insofar as they allowed for a longitudinal vibration as well as a transverse one. Cauchy overcame this incompatibility by considering the ether to be capable of changing to negative comprehensibility as well as positive comprehensibility. This negative compression allows the ether to react differently to various kinds of waves and to even allow the longitudinal velocity to be zero, for identifying as standing waves.

George Green (1793–1841) investigated Cauchy's results, and he found them to be inconsistent with conservation of kinetic energy. They were then criticized by Simeon Denis Poisson (1781–1840) and Franz Ernst Neumann (1798–1893) as being inconsistent with a wave theory that was developed more completely by Green.

Electromagnetic Rotation

As to how planets and other objects move through the ether, ideas came forth. Gabriel Stokes (1819–1903) suggested the effect is relative. The ether relates to slow-moving planets as a rarefied fluid, or jelly, and to the extremely rapid vibrations of light as a solid. James Mac Cullagh (1809–1847) proposed ethereal vortexes, or atoms, do not resist a displacement resulting in distortion of the medium; they change instead in their state of rotation. This process is of a transverse nature allowing atoms to move freely through ether with rotations subject to luminous effect.

With these two ideas combined, matter moves through ether without resistance, as if sinking into jelly, whereas light occurs as the changes in the rotational states of atomic-like vortexes. As rotation allows movement in the plenum, complex variation of rotation allows unlimited effects of light, electricity and magnetism from different types of wave action. Transverse wave polarization, for instance, could be a splitting of electric charge into positive and negative components.

Such speculative ideas were followed by empirical discoveries. In 1820, Han Christian Oersted (1777–1851) tested the effects of a magnetic needle near an electric current induced in a wire from a battery. He discovered the wire deflected the needle, as to reveal a connection between electricity and magnetism. Francois Arago (1786–1853) then discovered electric current magnetizes iron. Andre-Marie Ampere (1775–1836) then demonstrated that electric currents affect each other similar to the attraction and repulsion of magnetic poles. Electric currents repel each other when flowing in opposite directions; they attract each other when flowing in the same direction.

Accordingly, an electric current through a wire is assumed to emit virtual particles having more momenta in the direction of flow than perpendicular to it. With two wires parallel to each other and having their currents flowing in opposite directions to each other, the emitted virtual particles, in turn, collide to emit secondary virtual particles with more energy for repulsive effect. To the contrary, if the currents flow in the same direction, there is less secondary energy from the collision of virtual particles.

In 1831, Michael Faraday (1791–1867) discovered change in a magnetic field induces an electric current in a wire. It is thus only necessary to apply such a force as wind to move poles of a magnet to produce an electric current in a coiled wire that further produces additional magnetic effect. As an electric current produces an electromagnet, the electromagnet, in turn, produces more current. Alternating poles of the electromagnet being near the wire thus results as electromagnetic induction. It is how a generator is able to transform mechanical work into electricity.

A law to equate an electrical current to a magnetic field was developed by Jean-Baptiste Biot (1774–1862), Felix Savart (1791–1841) and Pierre Simon Laplace (1749–1827). A significant part of the law is a constant of proportionality c for equating a unit of electric charge e per time t passing through a unit length d of a section of wire in proportion to the magnetic pole strength p as $cp/e = d/t$. The pole strength p of the magnetic field is of the same dimensions as a unit of charge e of an electric field. They cancel each other out in the equation for c to be identified as speed $c = d/t$.

Wilhelm Eduard Weber (1804–1891) and Rudolph Kohlrausch (1809–1858) ascertained in 1836 a value of c as light speed being about 3×10^{10} centimeters per second.

The constant of proportionality c having the dimensions of a velocity was significant for the formulation of electromagnetic theory. Ampere had believed magnets are particular parts of electromagnets induced by electric currents within molecules of matter instead of within wires. However, Faraday believed magnetic currents, or "lines of force" in his way of thinking, exist in virtually quasi empty space whereby changes occurring in electromagnetic fields take time. Propagation of their effect is thus the propagation of light.

Faraday did not formulate a mathematical theory. His ideas along with others were included in a theory of electromagnetism formulated by James Clerk Maxwell (1831–1879). According to it, material wires are not needed to conduct electricity. It is able to propagate in the continuum of space as provided by the presence of an electromagnetic field alone. A displacement of electric current simply produces an electric field that induces a magnetic field that, in turn, induces another electric field for continuation of effect. An open field creation thus progresses at light speed as the electromagnetic spectrum.

Included in Maxwell's formulation of electromagnetism are Coulomb's law and the Biot-Savart law. An inverse square law for electrical force had been proposed by Joseph Priestley (1733–1804) and others. It was published first by Charles-Augustin de Coulomb (1736–1806). Similarly, in relation to a current of charge is the Biot-Savart law proposed by Felix Savart (1791–1862) and Jean Baptiste Biot (1774–1862) whereby the magnetic intensity between any two points along two different parallel wires is proportional to the distance squared between them and the amount of current flow and their speeds. If the currents flow is in the same direction, they attract; if they flow in opposite directions, they repel.

The two laws are united by Maxwell's theory of electromagnetism by a relationship of magnetic permissibility and electric permittivity. Constants of magnetic permissibility and electrical permittivity are denoted as μ and ε, respectively. Along with light speed c, they are c_0, μ_0 and ε_0 in vacuum space, as absent of matter, equating in the manner

$$c_0 = \frac{1}{\sqrt{\mu_0 \varepsilon_0}}$$

$$c_0^2 \mu_0 \varepsilon_0 = 1$$

It is with the presence of mass whereby the above constants have empirical effect.

Advancing Theory

A verification of Maxwell's theory of electromagnetism came in 1888 when Henrock Rudolf Hertz (1857–1894) produced electromagnetic waves interfering with themselves. He measured their wavelengths as fringes produced on a screen.

The next task became to determine the state of the ether in relation to matter. For instance, if light speed c is constant through ether, it is also consequential whether ether itself is in a state of motion that can or cannot influence the presence of matter.

Three possibilities were considered: (1) matter carries no ether, (2) matter partly carries it, and (3) matter carries all of it. Hertz adopted the third possibility, which had already been presented in 1845 by Stokes, but it was also inconsistent with experimental results. More consistent with experiment was a partial drag hypothesis proposed by Fresnel. A theory even more consistent with experimental results in general was one offered by Hendrick Antoon Lorentz (1853–1928) that assumes ether, as a state independent of matter, is a particular state of absolute rest. Although the assumption was not verified, further explanation led to the special relativity principle, which can explain all experimental results by assuming the properties of matter are altered to move through an invisible ether. Clocks moving through it, for instance, become slower, and matter becomes shorter in the direction of motion.

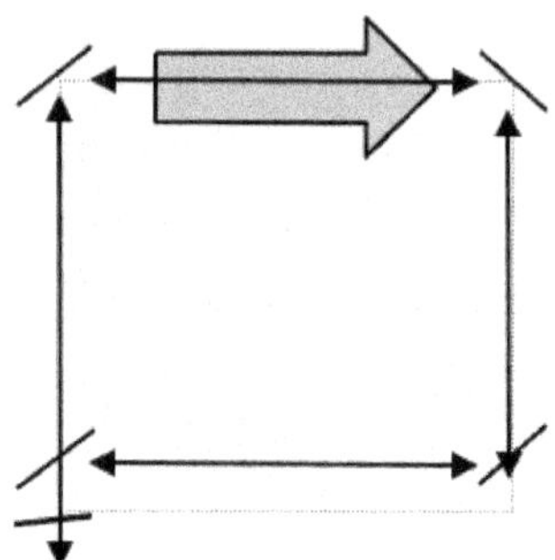

A particular experiment of significance at the time, illustrated above, was one suggested by Francis Arago indicating matter drags the ether. As light rays are split by a partially coated silver mirror, other mirrors direct the split parts for them to pass in opposite directions through a glass tube of running water. Rays moving in the same direction the water moves arrive back sooner than the rays moving in an opposite direction. However, the increased speed of some light and the decreased speed of other light are only equal to a fraction of the water's speed, as in agreement with Fresnel's partial drag motion.

The experiment supported Fresnel's partial drag hypothesis, but it was not able to explain another experiment by Edward Williams Morley (1838–1923) and Albert Abraham Michelson (1852–1931) in 1881, that did not include running water. If ether is dragged by matter, then a slight shift in fringes of the wave pattern on a screen should indicate a change in direction of Earth as it rotates on its axis and revolves around the sun. To the contrary, no appreciable change was observed to occur.

Explaining Experiment

The result of one experiment was consistent with Fresnel's partial drag hypothesis. Another experiment favored the no drag hypothesis. A more complete theory was needed in order for explanation of experimental results to be consistent. A near candidate at the time was Lorentz's electron theory.

Lorentz had theorized electromagnetic fields contain minute particles that bond together by their mutual attraction from having opposite charge. They thus exist within a state of equilibrium as a mutual state of vibration. Such a state affects light propagating through it. Because an electromagnetic field produces light, the state of the field itself is also changed. As the electromagnetic waves of light interfere with the equilibrium state of charged particles, the field reacts to a changed vibrant state of charged particles. The resulting effect is that of nullification of opposite factors for the ether to be invisible, and for matter of non-opposite factors to have visible effect on light propagation.

More specifically, to explain the null results of experiment, Lorentz assumed that matter moving through ether is contracted in the direction of relative motion. The contracted length of the apparatus in the relative direction of motion results in light moving the same distance it moves sideways.

Although contraction of length is sufficient by itself to explain the null results of the Michelson-Morley experiment, Lorentz went a step further in defining a local time in assuming clocks are similarly affected by moving through ether. Hence, a time of propagation is the same as if the apparatus is at absolute rest with the ether, suggesting a state of absolute rest in the ether is not observable as such. It need not be observable as such, according to the result of the Michelson-Morley experiment, because it can be any inertial state of motion instead.

Jules Henri Poincare (1854–1912) urged Lorentz in 1900 to generalize his theory to comply with the principle of relativity wherefore all motion of mass is merely relative, having no observable state of absolute rest. Lorentz did provide transformation equations to this effect, but he maintained there might still be a possible means of detecting a state of absolute rest. Meanwhile, Einstein independently derived the same Lorentz transformations in consistent manner of the ether allowing the relativity of motion through it.

Lorentz and Einstein's Explanations

Woldemar Voight (1850–1919) had derived these same transformation equations in 1887 to describe a rigid medium of light as a non-compressible fluid. Voight considered the theories of Cauchy, Neumann and others in view of a Doppler effect propounded by Christian Doppler (1803–1853). By it, sound waves or light waves are either stretched or contracted depending on the relative direction of motion between the observer and the source of emission. However, neither Lorentz nor Voight applied the transformation equations more generally to the laws of mechanics. Einstein did as such. As a consequence of it and the development of quantum physics, the ether as a methodological part of scientific theory became discarded.

In reformulating the principle of relative motion, Einstein emphasized the ether is unnecessary for formulating theory. He demonstrated this claim in 1905 with his formulation of special relativity theory by only postulating constant light speed with no reference to the ether as being empirically determined.

Einstein did suggest the ether could be used for understanding theory, just not being needed for its formulation, but other physicists did not agree it should be retained. With discoveries of both light and matter having dual particle-like and wave-like properties, such leading physicists as Bohr, Born and Heisenberg reinterpreted so-called wave equations as probability equations instead in determining probable location, time, energy and momentum for a particle effect to occur.

6

SIMPLE SPACE-TIME RELATIVITY

Einstein formulated special relativity (SRT) according to two postulates: (1) light speed is constant in a vacuum relative to the observer regardless of what the velocity of the observer is relative to any other and (2) physics laws are the same in all inertial reference frames. The second postulate is referred to as covariance. Light speed is not only covariant but is also particularly unique because of its speed in a gravitational free space being perceived the same by observers in inertial reference frames of relative motion.

How these postulates modify Galilean relativity and Newton's absolute space and absolute time according to constant light speed and the relativity of space-time is explained here according to the null result of the Michelson-Morley experiment. It continues with derivations of Lorentz transformations and so forth explained in view of modification of Galilean relativity and how observation of absolute rest is nullified.

The Michelson-Morley Experiment

The Michelson-Morley experiment was an attempted determination of light speed in relation to the ether. If the ether represents absolute rest and if light waves propagate at the same speed through it, then light speed should vary with Earth's motion varying through the ether in various orbital directions around the sun. However, no significant variation of light speed was found convincing enough by this experiment or any other. Null results were thus explained according to relative space-time instead of absolute time and absolute space.

The experimental measure of light speed was according to a to-there-and-back event. A silver-coated lens is positioned at a forty-five-degree angle to the incident light to split it into separate rays. One ray continued on through the lenses in the same direction while the other ray reflected at a right angle. Other lenses reflected the rays back to the silver-coated lens for them to pass through it and superimpose onto a screen.

What physicists expected from this arrangement was the total distance of each split ray moves differs due to perpendicular length of light movement differing from that of straight ahead and back movement. What actually appeared on the screen was an interference pattern of light and dark fringes (as predicted of superimposed waves). However, an appreciable amount of change in the interference pattern to verify change in light speed per perpendicular distance extended due to relative motion was what was not verifiably observed.

To explain the null results, Lorentz assumed length s_x of the apparatus contracts in the direction of relative motion by the factor

$$\alpha = \sqrt{1 - \beta^2}$$

β represents v/c as the direction and speed of matter in ratio to light speed. Perpendicular lengths s_x and s_y are considered equal:

$$s_x = s_y = s$$

Each arm of the apparatus compares as equal to a proper length denoted as s such that if the apparatus is theoretically at absolute rest, then respective times for light to move respective distances to-there-and-back are according to the following equations whereby x and y are perpendicular lengths extending from the observer's origin

$$2t_x = t_{x1} + t_{x2} = \frac{s}{c} + \frac{s}{c} = \frac{2s}{c}$$

$$2t_y = t_{y1} + t_{y2} = \frac{s}{c} + \frac{s}{c} = \frac{2s}{c}$$

In these equations, t_{x1} and t_{y1} are the respective times for light to move from the observer, and t_{x2} and t_{y2} are the respective times for light to be reflected back to the observer.

What if the apparatus moves at velocity v in the x direction relative to absolute rest?

According to Galilean relativity and the premise light speed is invariant relative to the ether, the times for light to move respective distances s_x and s_y vary. The differences in light speed and the apparatus speed are thus $c - v$ in the direction of motion and $c + v$ in the opposite direction of motion. A total time of the to-and-from propagation along the x-axis should be

$$2t'_x = t'_{x1} + t'_{x2} = \frac{s}{c - v} + \frac{s}{c + v}$$

$$= \frac{s(c+v)}{c^2 - v^2} + \frac{s(c-v)}{c^2 - v^2} = \frac{2sc}{c^2\left(1 - v^2/c^2\right)} = \frac{2s}{c\alpha^2}$$

This time differs from that of the apparatus being at absolute rest: $2t_x = 2s/c$. The ratio is per $\alpha^2 = 1 - v^2/c^2$. However, if the apparatus contracts in the direction of relative motion by the factor α, then the difference in time from that of absolute rest is only by the factor $1/\alpha$.

The total time it takes light to propagate in the perpendicular direction of motion, as during time $2t'_y$, is similarly determined. The actual path of light is according to two directions: (1) the direction along the arm of the apparatus perpendicular to the direction of motion and (2) the direction of motion in keeping pace with the apparatus. The actual distance is along the hypotenuse of a right triangle with respect to distance moved of those two other perpendicular directions. By the Pythagorean theorem

$$\left(ct'_y\right)^2 = s^2 + \left(vt'_y\right)^2$$

$$s^2 = \left(ct'_y\right)^2 - \left(vt'_y\right)^2$$

$$s^2 = \left(t'_y\right)^2\left(c^2 - v^2\right)$$

$$\left(t'_y\right)^2 = \frac{s^2}{c^2 - v^2} = \frac{s^2}{c^2(1 - \beta^2)}$$

$$t'_y = \frac{s}{c\alpha}$$

The time light moves to and from in the perpendicular directions is

$$2t'_y = \frac{2s}{c\alpha}$$

It is the same with respect to the direction of motion if the apparatus arms in the direction of motion contract by the factor α, being the square root of $(1 - v^2/c^2)$.

Because light moves at the same time, speed and distance along both arms of the apparatus, there are no differences observed of a shift in the pattern of fringes. Although this explains the null result of experiment, it only requires the apparatus to contract in the direction of relative motion. It does not require clocks to retard as well, and there remains the possibility of timing light from one place to another by synchronization of clocks by one being transported to the other location. Relative time is thus another step for determining spacetime relativity.

If clocks are slower by the factor $1/α$, duration of events are shorter by the factor α. Hence

$$2t'_x\alpha = 2t'_y\alpha = \frac{2s}{c\alpha}\alpha = \frac{2s}{c} = 2t_x = 2t_y$$

Absolute rest is thus invisible with regard to a to-there-and-back distance of light speed. However, to be consistent, a direct measure of light speed as constant should be nullified as well.

Further analyses involve more intricate explanations in relating such principles as covariance and the relativity of simultaneity. The latter explains how constant light speed is still determined as such from the synchronization of clocks by one of them being transported to a distance for its direct one-way measure.

Covariance

Covariance means the laws of physics apply the same to all systems. In other words, the perceptions of observers **A** and **B** are the same with regard to **A** or **B** being relatively at rest and the other moving at speed v. Because motion affects lengths and clocks, it is not obvious the speeds of different observers are perceived the same relative to each other, but relative motion is easily verified mathematically as covariant.

Consider observer **A** at absolute rest sees observer **B** approaching at velocity v the distance from x_1 to x_0 during time $t_1 - t_0$. The event includes the time it takes light to move a distance between positions x_1 and $x_0 = 0$. The time t_a of observer **A** seeing observer **B** moving the distance $x_2 - x_1$ is

$$t_a = \frac{x_1 - x_0}{v} - \frac{x_1 - x_0}{c}$$

Since the clock of observer **B** is slower by the factor $1/\alpha$, since observer **B** approaches the oncoming light at velocity v, and since coordinate lengths of observer **B** are relatively shorter by the factor α, the duration of observer **B** seeing observer **A** move along length $(x_1 - x_0)\alpha$, according to observer **B**'s clock, is

$$t_b = \frac{(x_1 - x_0)\alpha^2}{v} - \frac{(x_1 - x_0)\alpha^2}{c+v}$$

The task is to show $t_a = t_b$:

$$\frac{x_1 - x_0}{v} - \frac{x_1 - x_0}{c} = \frac{(x_1 - x_0)\alpha^2}{v} - \frac{(x_1 - x_0)\alpha^2}{c+v}$$

$$\frac{1}{v} - \frac{1}{c} = \frac{\alpha^2}{v} - \frac{\alpha^2}{c+v}$$

$$c(c + v) - v(c + v) = c\alpha^2(c + v) - vc\alpha$$

$$c^2 + vc - vc - v^2 = c^2\alpha^2 + vc\alpha^2 - vc\alpha$$

$$c^2 - v^2 = c^2\alpha^2$$

$$c^2 \left(1 - \frac{v^2}{c^2} \right) = c^2\alpha^2$$

Since $\alpha^2 = (1 - v^2/c^2)$, the equality $t_a = t_b$ is confirmed. Observers **A** and **B** thus perceive relative motion the same as if either A or B is at absolute rest.

Simultaneity

Even though Einstein postulated constant light speed as observable fact for no need of explanation, his principle of simultaneity helps explain it. Because observer **B**'s clock in relative motion is slow, it is not simultaneous with the clock of observer **A** relatively at rest. However, by the principle of covariance, **A** and **B** perceive events the same such that **B** is also regarded as relatively at rest. Simultaneity of events should thus be the same for both observers.

Consider **B** moves away from **A** at velocity v to a distance x after time t according to **A**'s clock. The task is to confirm **B**'s perception of time is the same as **A**'s. At distance x, it takes time x/c for light to move from **B** to **A** according to **A**'s spacetime coordinates. The time for **A** to see **B** move the distance x is thus x/v + x/c. However, coordinate lengths of **B** contract in the direction of motion by the factor α and duration of events by **B**'s slower clock are also shorter by the factor α. Because the difference in **B**'s velocity from that of light is c − v in relation to **A** being relatively at rest, a ratio of unity of respective times T_A and T_B of **A** 's and **B**'s clocks being synchronous in timing of the event is

$$\frac{T_B}{T_A} = \frac{\frac{x\alpha^2}{v}+\frac{x\alpha^2}{c-v}}{\frac{x}{v}+\frac{x}{c}} = \frac{\frac{\alpha^2}{v}+\frac{\alpha^2}{c-v}}{\frac{1}{v}+\frac{1}{c}} = \frac{\frac{\alpha^2(c-v)+\alpha^2 v}{v(c-v)}}{\frac{c+v}{vc}} =$$

$$\frac{\alpha^2 vc(c-v)+\alpha^2 v^2 c}{v(c-v)(c+v)} = \frac{\alpha^2 c(c-v)+\alpha^2 vc}{c^2-v^2}$$

$$\frac{\alpha^2 c(c-v)+\alpha^2 vc}{c^2\alpha^2} = \frac{c-v+v}{c} = 1$$

Relativity of simultaneity is thus covariant with respect to different speeds v and locations x for synchronization of transport of clocks and the direct measure of light speed.

Lorentz Transformations

Transformation equations transform distance and time coordinates of observers in relative motion to the coordinates of the other observer. Let x, y, z, t be the space-time coordinates of observer **A** relatively at rest, and let the corresponding coordinates of observer **B** relatively in motion be x', y', z', t'. Respective origins O and O' of the two coordinate systems overlap at time t = t' = 0, such that time and distance of the events in coordinate system **B**, as perceived by observer **B**, transform in view of observer **A**'s perception and vice versa for perceptions of each other's coordinates to be the same.

First consider a Galilean transformation of coordinates with regard to system **B** moving at velocity v relative to system **A**. Let origins of respective observers **A** and **B** be at the same place during the instant $t_0 = 0$ such that the coordinate distance x that system **B** moves at velocity v relative to **A** becomes shorter after time t by the amount vt. Hence

$$s = x - vt$$

The task is to show how observer **B** perceives the same rate of decrease in coordinate system s'.

The event could be in any direction of relative motion. To describe it according to Galilean relativity whereby clocks and distance coordinates do not appear affected by relative motion, three-dimensional-perpendicular coordinate systems are established. Coordinate system **B** has perpendicular coordinates from origin O at time t as x', y', z', t' in comparison to coordinates x, y, z, t for **A**, designating **B** in relative motion and A relatively at rest.

Lengths y' and z' are perpendicular to the direction of motion and do not contract by it. Hence, y' = y and z' = z. In the manner of explaining null results of the Michelson-Morley experiment, the distance light actually moves perpendicular to the direction of relative motion increases by the same mathematical factor the time of a clock in relative motion increases. The extended time it takes light to propagate farther is thus negated by the slower clock.

In contrast, the x and x' directions of relative motion moving relative to each other in opposite directions are consequential to how distances are determined by each observer. Since observer **B**'s clock is slow, **B** perceives a shorter duration of movement except for light from **A** reaching **B** sooner than light from **B** reaching **A**. Modification of Galilean transformations for determining distance and time coordinates is according to the equations

$$x' = \frac{x - vt}{\sqrt{1 - \beta^2}}$$

The symbol β represents the fraction v/c.

As for the transformation process, distances coordinates x' is the distance x minus distance vt in ratio to the relativistic factor due to **B**'s slower clock determining the event in comparison to **A**'s clock. With constant light speed as a measure of distance, the respective times t and t' can be replaced with x/c and x'/c.

$$t' = \frac{t - \frac{vx}{c^2}}{\sqrt{1 - \beta^2}}$$

Transformations generally apply to different perceptions of events instead of merely to relative motion between observers. Covariance still applies, but another observer or object in relative motion can be included in the analysis whereby its motion, time and distance are determined differently by two other observers even though the different transformation results still apply as though any inertial frame of motion can be considered as relatively at rest. The difference of how observers in different frames of reference determine the velocity of another is thus another aspect of consideration. It is referred to as an addition of velocities theorem.

Adding Velocities

Since clocks in relative motion differ, it is not obvious how observers in relative motion determine other velocities. Their calculation requires a theorem for adding velocities. For deriving it, consider inertial systems **A**, **B** and **C**. Consider **A** is relatively at rest, **B** as moving at velocity v_1 relative to **A**, and **C** as moving at velocity $v_2 = x'/t'$ relative to **B**. The sum of velocities v_1 and v_2 is to be determined relative to **A** as velocity $v_{12} = x/t$.

In relating coordinates of **B** to those of **A**, velocity v_2 transforms in the manner

$$v_2 = \frac{x'}{t'} = \frac{[x - v_1 t]\sqrt{1 - \frac{v_1^2}{c^2}}}{\left[t - \frac{v_1 x}{c^2}\right]\sqrt{1 - \frac{v_1^2}{c^2}}} = \frac{x - v_1 t}{t - \frac{v_1 x}{c^2}}$$

Multiplying the first and last sides of the equation by $t - v_1 x/c^2$ obtains

$$v_2\left(t - \frac{v_1 x}{c^2}\right) = x - v_1 t$$

$$v_2 t - \frac{v_1 v_2 x}{c^2} = x - v_1 t$$

Adding $v_1 v_2 x/c^2$ plus $v_1 t$ to both sides of the equation obtain

$$v_1 t + v_2 t = x + \frac{v_1 v_2 x}{c^2}$$

$$t(v_1 + v_2) = x\left(1 + \frac{v_1 v_2}{c^2}\right)$$

Finally, dividing both sides of the equation by t and $1 + v_1 v_2/c^2$ obtains

$$\frac{v_1 + v_2}{1 + \frac{v_1 v_2}{c^2}} = \frac{x}{t} = v_{12}$$

This equality represents velocity of system C in relation to the coordinate system A.

The formula above is for comparing velocities in the same direction of motion. It is possible to derive a formula for systems moving perpendicular to each other as well, which is actually simpler because of no contraction of length in the perpendicular direction of relative motion. Consider an object rotating perpendicular to the forward direction of relative motion. Because the clock in relative motion is slow, the observer perceives a faster speed of rotation by a relativistic factor. However, the actual speed of rotation is the combined vector product of perpendicular and forward speeds according to the Pythagorean theorem.

Consider **A** is relatively at rest, **B** is moving along the x-axis at velocity v_1, and **D** is moving along the y'-axis at velocity v_3. In relating coordinates of **B** to those of **A**, velocity v_3 becomes

$$v_3 = \frac{y'}{t'} = \frac{y\sqrt{1 - \frac{v_1^2}{c^2}}}{t - \frac{v_1 x}{c^2}} = \frac{y\sqrt{1 - \frac{v_1^2}{c^2}}}{t\left(1 - \frac{v_1 x}{tc^2}\right)}$$

Multiplying by $(1 - v_1 x/tc^2)$ and dividing by the square root of $(1 - v_1^2/c^2)$, the left and right sides of the equation become

$$\frac{v_3\left(1 - \frac{v_1 x}{tc^2}\right)}{\sqrt{1 - \frac{v_1^2}{c^2}}} = \frac{y}{t}$$

Since $x/t = v_1$, y/t becomes

$$\frac{y}{t} = \frac{v_3\left(1 - \frac{v_1 x}{tc^2}\right)}{\sqrt{1 - \frac{v_1^2}{c^2}}} = \frac{v_3\left(1 - \frac{v_1^2}{c^2}\right)}{\sqrt{1 - \frac{v_1^2}{c^2}}} = v_3\sqrt{1 - \frac{v_1^2}{c^2}}$$

This result is only of **D** moving along the y-axis relative to **A**. The velocity of **D** in the actual direction of the xy-plane has not yet been determined. To determine it, the Pythagorean theorem applies with regard to the speeds in perpendicular directions. The result is

$$v_{13} = \sqrt{\frac{y^2}{t^2} + v_1^2} = \sqrt{v_3^2\left(1 - \frac{v_1^2}{c^2}\right) + v_1^2} = \sqrt{v_3^2\alpha_1^2 + v_1^2}$$

The velocity v_{13} is that of system **D** relative to system **A**.

The derivation assumes light speed is the same for all observers, and they are to be consistent in showing light speed added to light speed or any velocity is still light speed c:

$$c_{cc} = \frac{c+c}{1+\frac{cc}{c^2}} = \frac{2c}{1+1} = c$$

$$v_{12} = \frac{v_1+c}{1+\frac{v_1 c}{c^2}} = \frac{v_1+c}{1+\frac{v_1}{c}} = \frac{c(v_1+c)}{c+v_1} = c$$

$$v_{12} = \frac{c+v_2}{1+\frac{cv_2}{c^2}} = \frac{c+v_2}{1+\frac{v_2}{c}} = \frac{c(c+v_2)}{c+v_2} = c$$

$$v_{13} = \sqrt{v_3^2\left(1 - \frac{c^2}{c^2}\right) + c^2} = c$$

$$v_{13} = \sqrt{c^2\left(1 - \frac{v_1^2}{c^2}\right) + v_1^2} = \sqrt{c^2 - v_1^2 + v_1^2} = c$$

Light speed is thus the same in all inertial systems according to the addition of velocities formulas.

Constant Speed Change

Acceleration of constant speed change increases without limit, but the addition of velocities theorem stipulates light speed is a limit for matter to neither exceed nor even reach. Applying the theorem to a system constantly changing speed is complex. Change in speed from one second to another can simply be determined according to the addition of velocities theorem for each second, but additional time units of additional slowing involve more entailed calculation.

Constant speed change is according to covariance in particular ways. Consider, as illustrated below, five systems **C, P', B, P, A**. **B** moves at velocity v relative to **A**. **C** moves at velocity v relative to **B** and at velocity v_{12} relative to **A**. **P'** decelerates from velocity v_{12} to velocities v and 0, being first relatively at rest with **B** and then relatively at rest with **A**. Opposite to deceleration is acceleration. **P** accelerates from O to v and v_{12}.

<pre>
C   v₁₂   ──────────────────────────────▶
P'   v₁₂   ───────────────▶ 0
B   v   ──────────▶
P   0   ──────────▶ v₁₂
A   0
</pre>

How does covariance apply?

Initial speed v is of **B** relative **A** and C, and of **P** and **P'** relative to **B**. They are covariant regarding A and **C** relative to each other and relative to **B**, and of B relative to P and P' and them relative to each other. Initial speed v_{12} of C relative to **A** is covariant of A relative to C along with covariance between **A** and **P'** and between **P** and **C**. Accelerations plus decelerations of **P** and P' are covariant relative to each other as well as relative to B, and vice versa.

Covariance maintains for each instant change in speed. More significant for continuation of theory is how each instant change maintains according to the combination of time and distance measure. Locally, slower time nullifies shorter distance. Non-locally, however, both the slower clock and shorter measuring rod relatively contribute to shorter distance. It is according to the relativistic factor squared, being consistent with the relativistic effects of gravity also being of constant acceleration.

Significantly, if v_{12} is extremely small in comparison to light speed c for it to be a negligible effect of the relativistic factor, such that it approximates as 1, then the above result equates as a Newtonian non-relativistic one:

What can further be explained is constant speed in analogy to a relativistic condition of gravity according to general relativity, both of them relating to a relativistic factor-squared.

$$\frac{v_{12}}{\sqrt{1-\frac{v_{12}^2}{c^2}}} = \frac{(2v)}{\left[1+\frac{v^2}{c^2}\right]\sqrt{1-\frac{v_{12}^2}{c^2}}} = \frac{2v}{\sqrt{\left[1+\frac{v^2}{c^2}\right]^2 - \left[1+\frac{v^2}{c^2}\right]^2\left[\frac{(2v)}{\left[1+\frac{v^2}{c^2}\right]c}\right]^2}}$$

$$\frac{2v}{\sqrt{1+\frac{2v^2}{c^2}+\frac{v^4}{c^4}-\frac{4v^2}{c^2}}} = \frac{2v}{\sqrt{1-\frac{2v^2}{c^2}+\frac{v^4}{c^4}}} = \frac{2v}{\sqrt{\left[1-\frac{v^2}{c^2}\right]^2}} = \frac{2v}{1-\frac{v^2}{c^2}} = \frac{2v}{\alpha^2}$$

The progression of v_0 to v_{12}, as for twice v, is thus in ratio to the relativistic factor squared.

The significance of the relativistic factor squared regarding constant speed change is that the relativistic factor for gravitational acceleration is also squared. It does not determine the value of the gravitational constant, but it is analogically consistent inasmuch as covariance of a clock increasing at constant speed equals that of a clock experiencing a constant increase in gravitational strength. However, the analogy is not merely that of centripetal acceleration and rotational acceleration. The increase in gravitational potential is analogous to increase in speed, as the relativistic effect includes both gravity and relative motion in the gravitational field.

Note: A mathematical solution to comparing P's clock time to A's is not here provided, because it entails a more complex form of covariance regarding gravitational acceleration in analogy to non-covariance acceleration of relative motion that is illustrative of a clock paradox. The link between them is significant in that special and general relativity unite according to it, and it is explained in chapter 8 in accordance with a Schwarzschild metric. Its explanation is also analogous to a clock paradox of special relativity.

The Clock Paradox

Observers **A** and **B** in relative motion time the other observer's clock as slow, but not if **B** moves away from **A**, reverses direction and then returns to **A**. **B**'s clock is then determined as the slower one by direct comparison. This is the well-known clock paradox: both clocks being perceived slower than the other by condition of covariance, but only one clock is slower by means of direct comparison.

The paradox is explainable because the event is not symmetrical. The traveling observer changes direction; the stay-at-home observer does not. In order to confirm the clock paradox does not contradict theory, it is only necessary to show how observer **A**'s clock is slow if observer **A** accelerates instead of observer **B**. Instead of observer **B** changing direction to return to observer **A**, **A** merely accelerates to catch up with **B**. There is symmetry with regard to either **A** or **B** accelerating: **B** moving away from **A** at velocity v uses force to become relatively at rest with system **A** and uses more force to return to **A** at velocity −v; as to compare with **A** using the force to become relatively at rest with system **B** and using more force to move at velocity v_{12} in order to catch up with **B**.

To verify symmetry is conditional, consider **B** moves along the x-axis relative to **A** at velocity v before changing direction along the x-axis at time T to move at velocity −v instead of v. With regard to **A** as relatively at rest, whereby **B** moves relative to **A**, the comparison of time recorded by **A**'s and **B**'s clocks during the trip is

$$2T' = \frac{2T}{\alpha}$$

This longer time results from **B** moving an extended distance at speed v because of a slower clock.

The determination of time for the previous event is simple. However, for the event of **A** being relatively at rest and then moving to catch up with **B**, **A**'s clock keeps two different rates while **B**'s clock remains the same. By the addition of velocities theorem a change in velocity is not simply from v to 2v; it is instead from v to v_{12}, which involves a corresponding change in rate of clocks. This condition is thus more complex than the previous one. However, the complexity was previously determined in the section "Constant Speed Change" whereby distance and time of acceleration equate the same as the average speed consistent with the Newtonian formula d = vt = (½)at².

The task is nonetheless to determine total time for **A** to catch up with **B** results the same as 2T/α. In perspective, what is to be determined is

$$T + T' = \frac{2T}{a^2} = \frac{2T}{a}$$

T' is **A**'s time while accelerating from speed zero to v_{12}. Since **B**'s clock is slow by the relativistic factor, **B** perceives T' as T, which is the same as the time **B** moves relative to A.

To reconfirm, **B** moves relative to **A** in time T the distance X at velocity v_1. In order to catch up with **B**, **A** accelerates from relatively at rest to velocity v_{12}. The time T' it takes **A** to catch up with **B** is according to difference in speeds: $v_{12} - (v_1 - 0)$. According to **A**, the difference is v_2; according to **B**, the difference is v_1. Hence

$$v_{12} = \frac{v_1 + v_2}{1 + \beta_1 \beta_2} = \frac{2v}{1 + \beta^2}$$

$$T' = \frac{X}{v_{12} - v} = \frac{X}{\frac{2v}{1+\beta^2} \neg v} = \frac{X(1+\beta^2)}{2v - v(1+\beta^2)}$$

$$\frac{X(1+\beta^2)}{2v - v - v\beta^2} = \frac{X(1+\beta^2)}{v - v\beta^2} = \frac{X(1-\beta^2)}{v(\ )1 - \beta^2}$$

$$\frac{T(1+\beta^2)}{1-\beta^2} = \frac{T(1+\beta^2)}{\alpha^2}$$

The total time relative to **A** is

$$T + T' = T + \frac{T(1+\beta^2)}{\alpha^2} = \frac{T\alpha^2}{\alpha^2} + \frac{T(1+\beta^2)}{\alpha^2}$$

$$\frac{T(1-\beta^2)}{\alpha^2} + \frac{T(1+\beta^2)}{\alpha^2} = \frac{T - T\beta^2 + T + T\beta^2}{\alpha^2} = \frac{2T}{\alpha^2}$$

However, being T' is slower than T by the factor 1/α, the total time of the event according to **B** is

$$\frac{2T}{\alpha^2}\alpha = \frac{2T}{\alpha}$$

Determinations of slower times are thus the same for **A** and **B** with regard to symmetry of conditions.

7

THE RELATIVITY OF MASS AND LIGHT ENERGY

Relativistic effects are covariant between systems with them being conditional to inertial velocity. Covariance is more paradoxical with regard to acceleration, exemplified by the clock paradox. Another paradox relates to how internal energy is explained. According to special relativity, relative increase in mass occurs with a relative increase in speed. It is simply explained by collisions being partly elastic and partly inelastic. The inelastic part prevents mass itself from obtaining light speed. However, another paradox is evident regarding how internal energy is itself explained in the manner mc^2. Explanation is according to internal action between particles of mass, but reference to light speed squared, c^2, is contrary to a condition of the Standard model of physics that such bosons as photons of light, or gravitons of gravity, are massless.

Conservation of energy is at issue regarding inelastic collision occurring between two masses for them to remain in the same state of relative motion as relatively at rest. For, if collision is inelastic in nature, then motion stops unless it continues by becoming some other form of energy. Internal energy could be stored, for instance, as heat or molecular motion. By elastic collision the kinetic energy is maintained as relative motion of the masses. For a collision to remain inelastic whereby the same mass content of the system remains as it was before collision, the internal energy of motion caused by the collision is spent to maintain in a state of equilibrium with its environment. It could be spent either as internal motion of molecules of matter or as such radiant heat as light to maintain in a thermodynamic state of equilibrium. However, the massless condition of the Standard model of particle physics is still a consequential issue to be understood.

How can massless radiation reduce mass of inelastic collision?

It can also be asked, what actually determines whether collision is elastic or inelastic? Would collisions between two particles in vacuum space be elastic?

Possible answers are that the mass is not allowed to lower itself in temperature or that some form of radiant energy in otherwise vacuum space converts mass into another form that can be recycled by the ability of the mass to explode outward with more absorption of light energy.

A previous issue once addressed by Newton and other physicists was that of explaining how heat as the internal molecules of matter within an isolated container in empty space applies regarding conservation of mass and momentum. It was considered that molecules might not be able to directly surrender their motion to other molecules as heat when they are separated by empty space. However, there is still the possibility to consider of energy being absorbed and emitted as light, whether visible or not by eyesight. Although photons are massless, there is a Higgs mechanism whereby a Higgs field of energy converts particular photons into mass particles. Further requirement here is that conservation of mass of inelastic collision need initially convert mass into massive radiation in order to comply with a relativistic reduction of mass-energy.

There are various possibilities of such interaction as that of elastic collision being dependent on particular states of equilibrium. If the collision is momentarily inelastic, for instance, radiant energy of some other form could be emitted out into space in various directions for conservation of mass in compliance with the relativistic condition that mass relatively at rest is less than it is if in relative motion. However, such results should have detectable effect in view of normal states of equilibrium. Establishing the laws of such states as occurring in vacuum space is thus critical for discovering other sources of energy.

Since energy is conserved of masses in collision by them exchanging it, kinetic energy of relative motion also maintains in some other form. Einstein agreed. Mass according to theory is one of many possible forms of energy that can convert from one form to another.

Einstein equated the internal energy of mass as a product of mass and light speed squared in the manner

$$E = \frac{E_0}{\sqrt{1-\frac{v^2}{c^2}}} = mc^2 = \frac{m_0 c^2}{\sqrt{1-\frac{v^2}{c^2}}}$$

Mass m_0 is the rest mass of m moving at velocity v, and m_0c^2 is internal energy constitutive of rest mass. The increased relative mass m in relative motion minus the rest mass m_0 is identified as kinetic energy potential for relative motion in approximation to kinetic energy of Newtonian Mechanics. It implies the same mass containing more heat as internal relative motion compensates for mass becoming relatively at rest.

Einstein explained internal mass-energy as work energy being emitted back and forth. Regarding his explanation, he considered the action within the mass as part of its internal mechanism in relation to light speed. A particle mass m as part of the total mass M relatively at rest is emitted within from position A to position B. Mass M then recoils during times t of emission and reception the distance x = vt while particle m moves distance d = ct. Work is thus done in relations to distances x and during time t that equates part of the potential energy Mv^2 to internal energy mc^2.

The total momentum equates in the manner

$$Mv = \frac{E_o}{t}$$

$$v = \frac{E_o t}{Mc}$$

Displacement x of M equates in the manner

$$x = vt = \frac{E_o t}{Mc}$$

$$xM = dm_o$$

$$x = \frac{dm_o}{M} = \frac{E_o t}{Mc}$$

$$\frac{E_o t}{Mc} = \frac{dm_o}{M}$$

$$E_o = \frac{Mm_o cd}{Mt} = \frac{m_o cd}{t} = m_o c^2$$

Note: Momentum of emission and absorption of a photon is assumed to be conserved in this derivation. Its conservation is of an internal action of the system that does not change in relative mass. It is therefore in compliance with both Galilean relativity and Newtonian Mechanics.

However, the relativity of mass-energy is a lot more complex than that of Newtonian mechanics. According to relativity, for instance, elastic collision is not necessarily totally elastic. A transfer of mass from one to another occurs instead. Its occurrence prevents mass itself from accelerating to light speed, in being consistent with the addition of velocities theorem. Even the reflection of light is more complicated. Consider, for instance, a mass completely absorbing a photon for its momentum to equate to that of the photon. If the velocity becomes three-fifth light speed before mass reflects the photon in the opposite direction, then the mass should obtain six-fifth light speed regarding conservation of momentum, which is contrary to the principle of special relativity theory requiring mass can neither exceed nor even reach light speed. For consistency of theory, the condition is explainable as an increase in mass at faster speed for maintaining conservation of momentum at lesser speed than would be without an absorption of mass.

Although photons of light are massless according to today's Standard model of physics, analyzes of interactions between mass and light are here merely provided for simplicity of interaction being consistent with a particular particle nearly at light speed. Photon energy is consistent with the process and could, in part, be a contributor to it.

The following analyses includes such principles as the Doppler effect. It applies to how change in relative motion occurs from various ways of interaction between light and other mass particles. Further analyses include how conservation of momentum and energy apply of the particular case of mass absorbing and reflecting light according to both inelastic and elastic collision. More conservation of momentum and energy is then more generally explained according to how mass interacts with other mass.

Although the interaction of light and mass is explained in the same manner of the interaction of mass with other mass, it is merely such in assuming that it is according to an internal mechanism that is consistent with the relativity of covariance. A Higgs field capable of converting particular photons into mass particles is indicative of its possibility.

The Doppler Effect

Conservation of momentum regarding partial elastic and inelastic collision between mass and light is conditional to Doppler effect. It applies more generally to either interaction between photons and mass or between matter with other matter.

Although increase in mass-energy along with relative motion is complicated, it was previously indicated by a Doppler effect proposed in 1842 by Christian Doppler (1803–1853). It explains why a whistle on a train is of a lower pitch when the train recedes from the observer, and of a higher pitch when approaching the observer. Explanation is according to the vibration of air molecules forming waves. Lower pitch sound is described as stretched out weaker waves and higher pitched sound is more compacted stronger waves. There is stretching of longer wavelength resulting from the recession between the source and observer, and shorter wavelength observance results from the observer and source approaching each other.

Light energy being categorized as wave action has similar effect. Receding objects emit light that is received of longer wavelength and approaching objects emit light that is received of shorter wavelength. The shorter wavelength of light is more energetic similar to shorter wavelength of sound being more frequent and louder of higher pitch.

These effects are of the general dynamics of ordinary particles as well as for light. Bullets fired from a gun, for instance, are more energetic if the gun firing them is moving towards the target rather than away from it. They also apply to light whether light is either particle or wave in nature. Systems approaching each other naturally receive light signals more rapidly than do systems receding from each other. Such effects result from laws of motion partly according to Newtonian mechanics and more generally according to relativity theory.

The task is to illustrate general covariance of observers firing bullets at each other for comparison of results with a specific Doppler effect of light propagation. Accordingly, observer **A** is relatively at rest at the origin from where observer **B** moves away at velocity v_1. After time T and the distance X of their separation, as according to **A**, **A** fires a bullet at **B**. After time T' = T/α_1 of their separation, as according to **A**, **B** fires a bullet at **A**. Regarding covariance of the same relative speed, the time for receiving a bullet is the same for **A** as it is for **B** even though the bullets move away relatively faster than it does approaching with respect to either **A** or **B** being considered relatively at rest. Conservation of both momentum and energy thus applies according to the relativity of covariance that was previously analyzed in the previous section of this chapter with the exchange of photons.

First, consider a time T$_b$ for **B** to receive a bullet from **A** is time T plus time T$_x$ it takes the bullet moving at velocity v$_2$ to catch up with **B** moving away from X at velocity v$_1$. Hence

$$T_b = T + T_x = \frac{X}{v_1} + \frac{X}{v_2-v_1} = \frac{x(v_2-v_1)}{v_1(v_2-v_1)} + \frac{v_1 X}{v_1(v_2-v_1)}$$

$$\frac{X(v_2-v_1)+v_1 X}{v_1(v_2-v_1)} = \frac{v_2 X - v_1 X + v_1 X}{v_1(v_2-v_1)} = \frac{v_2 X}{v_1(v_2-v_1)}$$

This time is according to the clock of **A**. Because **B**'s clock is slower by the factor 1/α_1, as to perceive less duration, **B** determines the total time as

$$T_b = \frac{X\alpha_1}{v_1}\left[\frac{v_2}{v_2-v_1}\right]$$

The task now is to determine the time for **A** to receive a bullet from **B**. Because **B**'s clock is slow by the factor 1/α_1, the time **B** decides to shoot a bullet at **A** after **B** passes **A** is T' = T/α_1 for moving distance X' = x/α_1 between **A** and **B**. The bullet's speed is calculated according to the addition of velocities theorem whereby v$_{12}$ and v$_2$ are both negative values because of **B** receding from both **A** and the bullet:

$$-v_{12} = \frac{v_1+(-v_2)}{1+\beta_1(-\beta_2)} = \frac{v_1-v_2}{1-\beta_1\beta_2}$$

Although **A** receives the bullet as moving in the negative direction, as v$_{12}$ to be a negative velocity, the time X'/v$_{12}$ for **A** to receive the bullet from **B** is additional to the time X'/v$_1$. It is thus positive. Hence, the time **A** receives the bullet is

$$T_a = T' + T'_x = \frac{X'}{v_1} + \frac{X'}{v_{12}} = \frac{X}{v_1 \alpha_1} + \frac{X}{v_{12}\alpha_1} = \frac{v_{12}X}{v_1 v_{12}\alpha_1} + \frac{v_1 X}{v_1 v_{12}\alpha_1}$$

$$\frac{v_{12}X + v_1 X}{v_1 v_{12}\alpha_1} = \frac{X(v_{12}+v_1)(1-\beta_1\beta_2)}{v_1(v_2-v_1)\alpha_1} = \frac{X}{v_1\alpha_1}\left[\frac{v_{12}(1-\beta_1\beta_2)+v_1(1-\beta_1\beta_2)}{v_2-v_1}\right]$$

$$\frac{X}{v_1\alpha_1}\left[\frac{(v_2-v_1)+v_1(1-\beta_1\beta_2)}{v_2-v_1}\right] = \frac{X}{v_1\alpha_1}\left[\frac{v_2-v_1+v_1-v_1\beta_1\beta_2}{v_2-v_1}\right]$$

$$\frac{X}{v_1\alpha_1}\left[\frac{v_2-v_1\beta_1\beta_2}{v_2-v_1}\right] = \frac{X}{v_1\alpha_1}\left[\frac{v_2-v_2\beta_1^2}{v_2-v_1}\right] = \frac{Xv_2}{v_1\alpha_1}\left[\frac{1-\beta_1^2}{v_2-v_1}\right]$$

$$\frac{X\alpha_1^2}{v_1\alpha_1}\left[\frac{v_2}{v_2-v_1}\right] = \frac{X\alpha_1}{v_1}\left[\frac{v_2}{v_2-v_1}\right]$$

This time being the same as T_b, as perceived by **B,** thus complies with covariance.

A similar event for light signals is achieved by substituting c for v_2, and v for v_1, to obtain

$$\frac{X\alpha}{v}\left[\frac{c}{c-v}\right] = \frac{X\alpha}{v}\left[\frac{1}{1-\beta}\right] = \frac{X}{v}\left[\frac{\alpha}{1-\beta}\right] = \frac{X}{v}\left[\frac{\sqrt{1-\beta^2}}{1-\beta}\right]$$

$$\frac{X}{v}\left[\frac{\sqrt{1-\beta}\sqrt{1+\beta}}{1-\beta}\right] = \frac{X}{v}\left[\frac{\sqrt{1+\beta}}{\sqrt{1-\beta}}\right] = \frac{X}{v}\left[\frac{1+\beta}{1-\beta}\right]^{\frac{1}{2}}$$

This is the time either Observer **A** or Observer **B** sees the other observer move the distance X.

The relation holds for any distance X or X', and for any time T_a or T_b. It is thus divisible into any number of distances and times. These times and distances relate to wave properties of light as wavelength λ and frequency f such that

$$f' = f\left|\frac{1-\beta}{1+\beta}\right|^{\frac{1}{2}} \qquad \lambda' = \lambda\left|\frac{1+\beta}{1-\beta}\right|^{\frac{1}{2}}$$

$$f' = f\left|\frac{1+\beta}{1-\beta}\right|^{\frac{1}{2}} \qquad \lambda' = \lambda\left|\frac{1-\beta}{1+\beta}\right|^{\frac{1}{2}}$$

The equations are relativistic Doppler formula for comparing light sources and observers as relatively either approaching toward or receding from each other at constant velocity. The results are the same whether of light or sound waves or of particles of matter.

More general results are more complex regarding the interaction of light and mass. Light is now categorized as a photon of zero mass according to the Standard model of physics. However, previous analysis in this chapter seemingly indicates results to the contrary. A means of its verification would be an experiment producing momentum by difference in color. If black absorbs light and white reflects it, then sheets having opposite sides of black and white that are reversed on each side of a pole should be able to rotate the sheets around the pole. However, experiments have only shown absorption and reflection of light occur according to frequency instead of intensity. More electrons, for instance, are emitted back from a metal as such.

Massless light particles are also distinguishable from mass particles having wave properties. Massless wave properties differ from mass particles according to a Pauli exclusive principle. According to it, mass cannot occupy the same space whereas waves can superimpose to combine their presence to be able to pass through mass. The passage could even be undetected if in an equilibrium state whereby effects are nullified by other effects. Such an equilibrium state is consistent with the principles of entropy and covariance. By disorder of entropy, differences in the equilibrium states are how change occurs from their interaction.

Overall, outer-space contains starlight crossing paths from all directions from countless sources, but no interference of their crossing is indicated by general observation, except for new discoveries from more advanced telescopes transported into outer space.

Mass-Energy and Light

Conservation of energy is at issue regarding inelastic collision occurring between two masses for them to remain in the same state of relative motion as relatively at rest. For, if collision is inelastic in nature, then motion stops unless it continues as some other form of energy. Internal energy could be stored, for instance, as heat or molecular motion. By elastic collision, the kinetic energy is maintained as relative motion of the masses. For a collision to remain inelastic whereby the same mass content of the system remains the same as it was before impact, then the internal energy of motion caused by the collision is not spent for it to maintain as such in a state of equilibrium with its environment. By special relativity, the two masses becoming relatively at rest should be of less mass. However, mass collisions do result in the production of heat that can be released as a particular form of radiation similar to how mass can be created from light according to the Higgs mechanism. For conservation, mass released as massless radiation need be part of the Higgs mechanism producing massive bosons from such massless ones as photons with regard to a recycling process.

Heat is viewed as a random motion of internal molecules of mass, but conservation of relative mass and momentum applies to molecular motion. The molecules are not able to directly surrender their motion to molecules as heat when they are separated by empty space. However, there is still a possibility to consider of energy being absorbed and emitted as such electromagnetic radiation as light. If inelastic collision occurs between two masses, a decrease in mass occurs according to special relativity as observed from observers relatively at rest. If this event occurs in so-called empty space, then some other form of radiation having mass needs to be emitted instead of photons of no mass.

Since energy is conserved of masses in collision by them exchanging it, kinetic energy of relative motion also maintains in some other form. Einstein agreed. Mass according to theory is one of many possible forms of energy that can convert from one form to another.

Einstein equated the internal energy of mass as a product of mass and light speed squared in the manner

$$E = \frac{E_0}{\sqrt{1-\frac{v^2}{c^2}}} = mc^2 = \frac{m_0 c^2}{\sqrt{1-\frac{v^2}{c^2}}}$$

Mass m_0 is the rest mass of m moving at velocity v, and m_0c^2 is internal energy constitutive of rest mass. The increased relative mass m in relative motion minus the rest mass m_0 is identified as kinetic energy potential for relative motion in approximation to kinetic energy of Newtonian mechanics. However, there is a paradox with regard to a decrease in relative mass resulting from inelastic collision. For conservation of mass to comply, it needs to be maintained by internal motion being increased as kinetic heat energy.

Relative Mass Increase

According to special relativity, a mass m in relative motion at velocity v is relatively greater than m_0 relatively at rest:

$$m = \frac{m_0}{\sqrt{1 - \frac{v^2}{c^2}}}$$

Further according to theory, in general, mass-energy is conserved after collision. If the collision is inelastic whereby relative motion between masses is terminated, part of the total mass-energy could become another form of energy, such as heat or electromagnetic radiation. However, it is here assumed, for analytical purpose, to be maintained instead according to a special condition of equilibrium that it is part of.

An increase in relative mass along with an increase in relative motion is mathematically verifiable according to conservation principles of energy and momentum. Constant light speed and covariance apply in manner consistent with the addition of velocities theorem as well

Mass in relative motion is first distinguished apart from mass relatively at rest. Consider mass m_0 to be relatively at rest and mass m to be moving at velocity v_1. They become one mass $M = m + m_0$ by means of inelastic collision according to conservation of mass. and momentum, in the manner

$$mv_1 + m_0(0) = Mv_2 = mv_1$$

$$(m + m_0)v_2 = mv_1$$

$$mv_2 + m_0v_2 = mv_1$$

$$mv_2 - mv_1 = -m_0v_2$$

$$m(v_2 - v_1) = -m_0v_2$$

$$m = \frac{-m_0v_2}{v_2 - v_1} = \frac{m_0v_2}{v_1 - v_2}$$

The objective is to verify

$$m = \frac{m_0 v_2}{v_1 - v_2} = \frac{m_0 v_2}{\sqrt{1 - v_1^2/c^2}}$$

It proceeds in the manner

$$\frac{m}{m_0} = \frac{v_2}{v_1 - v_2} = \frac{1}{\sqrt{1 - v_1^2/c^2}}$$

$$\sqrt{1 - v_1^2/c^2} = \frac{v_1 - v_2}{v_2}$$

$$v_2 \sqrt{1 - v_1^2/c^2} = v_1 - v_2$$

$$v_2^2 \left(1 - v_1^2/c^2\right) = v_1^2 - 2v_1 v_2 + v_2^2$$

$$v_2^2 - \frac{v_1^2 v_2^2}{c^2} = v_1^2 - 2v_1 v_2 + v_2^2$$

$$-\frac{v_1^2 v_2^2}{c^2} = v_1^2 - 2v_1 v_2$$

$$2v_1 v_2 = v_1^2 + \frac{v_1^2 v_2^2}{c^2}$$

$$2v_1 v_2 = v_1^2 \left[1 + \frac{v_2^2}{c^2}\right]$$

$$2v_2 = v_1 \left[1 + \frac{v_2^2}{c^2}\right]$$

$$\frac{v_2 + v_2}{1 + \frac{v_2^2}{c^2}} = v_1$$

The last equation is of the form of the addition of velocities theorem. It relates to the relativistic increase in mass in manner of an elastic collision being the reverse of an inelastic one. Consider, for instance, $v_1 = (3/5)c$ whereby $m/m_0 = 5/4$ and numerical value of total momentum is $mv = (3/5)(5/4) = ¾$. After inelastic collision the total mass becomes $(5/4)m_0 + m_0 = (9/4)m0$. For conservation of momentum, the numerical values become again $(9/4)(1/3) = ¾$. By elastic collision, the moving mass simply becomes the previous rest mass and the previous rest mass becomes accelerated to $(3/5)c$ for the numerical value of total momentum to remain in accordance with the addition of velocities theorem as

$$[(1/3 +1/3)/(1 + 1/9)](5/4) = (3/5)(5/4) = ¾$$

Conserving Energy of Light and Mass

If light possessed momentum, its emission and absorption by mass would be subject to conservation laws of momentum and energy. Mathematical verification by example is with regard to mass $m_0 = 1$ and momentum $m_0 = 0$ relative to observer **A** absorbing a photon of light having momentum $m_x c$ such that m_0 and the photon energy move with observer **B** at velocity $.6c$ relative to observer **A**. Conservation of momentum with regard to total mass of both light and matter with respect to **A** is according to the equations

$$m_x c + m_0(0) = (m_x + m_0).6c$$

$$m_x c = .6m_x c + .6m_0 c$$

$$m_x = .6m_x + .6m_0$$

$$.4m_x = .6m_0$$

$$m_x = 1.5m_0$$

Total mass with respect to A before and after absorption is thus

$$1.5m_0 + m_0 = 2.5m_0$$

Total momentum before collision was $(1.5)(1) = 1.5$ mass-velocity units. It is $(1.5 + 1)(.6) = (2.5)(.6) = 1.5$ mass-velocity units after collision. The total momentum is thus conserved with regard to absorption of light by matter.

For a reverse process of inelastic collision becoming elastic collision, a photon is emitted from total mass $m_x + m_0$ in the opposite direction at which it was absorbed. Although **B** perceives speed of emission the same as absorption, **A** perceives the emitted photon speed as

$$v_{12} = \frac{2v}{1+\frac{v^2}{c^2}} = \frac{2(.6c)}{1+.36} = \frac{15}{17}c$$

The relative mass becomes

$$\frac{m_0}{\sqrt{1-\left[\frac{15}{17}\right]^2}} = \frac{17}{8}m_0$$

Its momentum becomes

$$\frac{17}{8}m_0 \cdot \frac{15}{17}c = \frac{15}{8}m_0c$$

The photon *mass-inertia* in the opposite direction is

$$m_{xb} = m_x \left|\frac{1-v_{12}}{1+v_{12}}\right|^{\frac{1}{2}} = \frac{3}{2}m_0 \left|\frac{c-\frac{15}{17}c}{c+\frac{15}{17}c}\right|^{\frac{1}{2}} = \frac{3}{8}m_0$$

Total mass and total momentum with respect to A are again

$$\frac{17}{8}m_0 + \frac{3}{8}m_0 = \frac{5}{2}m_0$$

$$\frac{17}{8}m_o\frac{15}{17}c - \frac{3}{8}m_0c = \frac{3}{2}m_0c$$

Both mass and momentum of light and matter are thus conserved of partial elastic collision between light and matter by means of a transfer of mass between light and matter.

Modifying Relativity

According to theory, internal energy is mathematized as mc² for any amount of mass m moving at any velocity v. Light speed squared is the same for any amount of mass, but kinetic force is variable for both m and v, and is (½)mv² instead of mv². Although it appears c can still be the upper limit, further calculation indicates more internal energy is lost than gained if v in relation to relative motion or gravitational potential exceeds one-half c.

Consider further how internal energy of mass interacts with itself as collisions between internal particles moving at light speed. With mass M moving at velocity v, internal particles moving at speed c collide with the forward parts of M at speed difference c minus v in the direction of relative motion but at speed difference c plus v in the opposite direction of its relative motion. As particle action, there is more kinetic energy pushing backward than forward. It should slow down contrary to conservation of momentum except for the back-and-forth transfer of momenta between the particle and mass M. According to Doppler effect, more energy transfers into the particle from backward collision than forward collision. There is thus also increase in particle energy for relatively more forward push and decrease in particle energy for less backward push.

M is considered a container with enough energy to not be pushed apart by elastic collision of the particle, and it is somehow able to maintain its state of relative motion if there is no outside influence to change it. Conservation of momentum maintains after each back-and-forth collision. For continual maintenance, a more complex interaction of the system as a whole somehow needs to occur. Internal mass-energy is thus more complex than that of a container moving through space with an object inside moving back and forth.

Still, how does interaction occur of internal mass-energy? Mass could also be moving through a medium of otherwise vacuum space according to some kind of wave action. Waves differ from particles inasmuch as particles cannot occupy the same space whereas waves can superimpose to pass on through. There is the Higgs mechanism, for instance, whereby light of zero mass can become massive. Ocean waves, for example, leave momentum of water the same after passing on over, but they could push something forward that is either in the water or on the shore.

How mass exists is itself questionable. It could be composed of waves superimposing as wave-packets. Their equilibrium states further depend on how they interact with the medium of space itself, whose existence is another questionable aspect of reality not fully understood as completely verifiable. It nonetheless allows for explanation.

Since waves themselves can be invisible moving through a medium, it is possible a medium itself composed of wave action is itself invisible. As to how it applies to conservation of momentum, superimposing of wave-packets depends on wave frequency. By the Doppler principle, waves and wave-packets approaching each other reflect each other with more energy. The shorter time of wave impact constitutes its more energy density in terms of frequency. Optional, however, is whether waves are reflected or allowed to superimpose and pass on through. It could be the decision is according to the packet's state of relative motion along with its internal interaction between particle-like waves. If mass M allows more frequent waves to superimpose while moving towards them, and allows lesser frequency of waves superimposing in the same direction of relative motion, internal and external interactions then balance out in manner of maintaining conservation of momentum.

Mass moving forward can thus interact with an outside medium in different ways than merely as particle interaction. However, the relativity of mass-energy is a lot more complex than that of Newtonian mechanics. According to relativity, for instance, elastic collision is not necessarily totally elastic. A transfer of mass from one to another occurs instead. Its occurrence prevents mass itself from accelerating to light speed in being consistent with the addition of velocities theorem. Even the reflection of light is more complicated. Consider, for instance, a mass completely absorbing a photon for its additional momentum to equate to that of the photon. If the velocity becomes three-fifth light speed before mass reflects the photon in the opposite direction, then the mass should obtain six-fifth light speed regarding conservation of momentum, which is contrary to the principle of special relativity theory requiring mass neither exceed nor even reach light speed. For consistency of theory, the condition is explainable as an increase in mass at faster speed for maintaining conservation of momentum at lesser speed than would be without absorption of mass.

Although photons of light are considered massless, explanation is merely according to particular aspects of wave action.

Waves themselves are considered massless insofar as their movement through mass only momentarily changes the internal composition of mass. After the waves pass through, the content of mass can return back to the same state it was in before encountering the wave. However, change in the state of mass can also occur. By a Compton effect, for instance, gamma rays cause radioactive decay of atomic nuclei. By Einstein's photoelectric effect, waves of enough frequency cause electrons to be emitted from a metal. For these effects to occur, waves themselves need to be comprised of a substance whether it is of ocean waves of water or ethereal waves of a general medium for all of mass.

By the Pauli exclusive principle, particles cannot occupy the same space whereas waves can superimpose to pass on through possibly with invisible effect. Assume mass consists of particular states of wave superposition as wave-packets. Further assume light and matter exist as two different forms of energy that can, according to particular conditions, convert from one form to another. Superimposing of light waves is thus conditional. For explaining a possible condition according to the relative motion of mass, assume it is a massive form of electromagnetic energy potential as wave-packets existing within an equilibrium state except for reflecting waves within a range of particular lengths and frequencies.

Certain conditions for either reflection or passing on through are consistent with relativity theory. If the wave-packet is relatively at rest, then what waves are reflected is according to how their potential energy frequencies change accordingly. More frequent waves reflected when the wave-packet is relatively at rest. When it is in relative motion, less frequent waves are reflected forward in the direction of relative motion and more frequent ones are reflected backward in the opposite direction of relative motion. Overall, the wave packet merely obtains a new state of equilibrium consistent with the relativity of motion.

More detailed explanation also includes, in view of the Doppler effect, frequency of shorter waves is of more dense energy per time.

The relativity of motion is only one circumstance. Others are of gravity and the quantum nature of the atom. Gravitons are also considered massless. How they can gravitate as such for change in momentum is more complex. Even more complexity occurs according to the atomic structure whereby the Higgs Mechanism of a Higgs Field somehow allows particular massless bosons of light energy to transform into massive ones.

How can massless gravitons continually change mass momenta? It is more complex. For mass evenly distributed throughout space, gravity nullifies itself. Gravitons have massive effect only in ratio to the uneven distributions of mass.

Consider a hollow container moving freely in space at ½ light speed. Inside it is a photon moving back and forth in the container's direction of motion. If the photon is absorbed by particles of the same mass as the photon, then in the same direction of motion they obtain velocity three-half c in accordance with conservation of momentum. If the photon is reflected back in the opposite direction, its momentum numerically becomes in ratio to light speed as (3/2 + 2/2) – 5/2. After reflection, the photon is then absorbed by a rear particle of the same mass. The numerical momentum of the rear particle becomes –½. After reflection, the momentum numerically becomes –3/2. Total particle momentum, after the process, thus stays at 5/2 – 3/2, being equal unity for twice mass moving at one-half lightspeed. As to why these two particles do not continue to move apart from each other is due to the hollow container containing them by means of its attractive-binding-energy.

Consider the container moving at 1/3 lightspeed. Numerical values of the forward particle become 4/3 + 3/3 = 7/3. Numerical values of the rear particle become –2/3 – 3/3 = –5/2. Total numerical momentum is 2/3, being a fractional difference less in comparison to the container's momentum as a whole, as for having more containment than needed.

Consider the container moving at 2/3 light speed. Numerical momentum values of the forward particle become 5/3 + 3/3 = 8/3. Numerical momentum values of the rear particle become –1/3 – 3/3 = –4/3. Total momentum is numerically 8/3 – 4/3 = 4/3, being a fractional difference greater in comparison to the container's momentum as a whole, as for having less containment than needed.

The former result being positive and the latter being negative indicates the one-half speed difference in between is a determining factor regarding mass and its upper light speed limit.

There is a similar calculation regarding gravitational-escape-speed in ratio to orbital speed. There is also an ethereal theory by Harold Aspen equating internal mass-energy to an electrostatic formula in relation to kinetic energy. It determines the ratio of the proton mass to electron mass, and it also derives the numerical value of the gravitational constant according to it. The results are according to loss of energy contributing to the process.

Some physicists have pointed out contradictions pertaining to the expanding universe theory and the black hole condition of general relativity. Such contradictions become explainable paradoxes according to a one-half lightspeed condition as the upper limit for physical containment.

8

THE RELATIVITY OF GRAVITY

After Einstein modified Galilean relativity for it to comply with relative spacetime instead of absolute space and absolute time, he focused on Newton's theory of gravity for its compliance with relative spacetime as well. He considered gravity as analogous to relative motion in compliance with an equivalence principle fundamental to both Newtonian mechanics and relativity theory whereby gravitational mass and inertial mass are essentially the same. A change in motion of mass thus occurs either from a collision with another mass or by gravity of the other mass. Twice as much mass changes velocity of another mass twice as much by either gravity or collision with the other mass.

Another principal Einstein used in relating gravity to relative motion is similar to one Copernicus previously proposed to explain our unawareness of Earth's orbital motion. Copernicus realized that we are not internally aware of Earth's motion around the sun because of us being in the same uniform motion with Earth around it. Einstein also reasoned we are internally unaware of falling freely by gravity because, as had been proposed by Galileo, all mass falls at the same rate through a vacuum towards Earth's center. Einstein also assumed light is gravitated along with mass for no internal awareness of change, which is consistent with describing spacetime according to constant light speed. However, gravity is inhomogeneous by nature to complicate the equivalence principle in that parts of a system gravitate towards a center of mass according to various distances and different directions of free fall.

Einstein equated free fall with inertial motion inasmuch as there is no internal awareness of either one, but free fall of Earth with its moon can be felt by way of ocean tides because of parts of Earth being closer to the moon gravitating more towards it. There is also a tendency of mass to gravitate towards a common center instead of in parallel direction. Objects inside a container in free fall thus tend to converge towards a mass center. Furthermore, what is opposite to falling inward is orbital speed. An increase in orbital speed at the same radial distance results in a straighter orbital path for moving farther contrary to falling at the same rate. A ball moving faster also moves slightly farther before lowering to the ground because of Earth's surface curvature. The greater speed of light perpendicular to the radial distance similarly moves straighter distances than does slower mass falling at the same rate towards the center of mass. There is thus a greater tendency of orbital escape by faster mass and light.

Since constant light speed is the founding principle of SRT and since light is assumed to gravitate toward matter, Einstein relegated the validity of SRT as a special case, applicable as far as tidal effects are negligible. They are negligible, for instance, inasmuch as particular parts of the gravitational field are perceivable as homogeneous whereby different distances are too short to determine their differences of gravitational effect. Einstein then opted for a geometrical description of gravity in accordance with spacetime curvature due to the presence of mass.

Even though Einstein opted to stipulate SRT is valid only as a special case, it can still be considered as an integral part of GRT inasmuch as the latter contains forms of acceleration whereby non-symmetrical conditions exist in analogy to the clock paradox. They interrelate by means of invariance.

Invariance

The Pythagorean theorem is useful in surveying when the line of sight for a direct measure of distance is blocked. A right triangle hypotenuse provides an invariant for mapping an alternative direction as a detour, with the hypotenuse as determinable for distance. Because its two other side-lengths can extend perpendicular to each other in any directions from opposite ends of the hypotenuse, their relation to the hypotenuse determines an invariant for all right triangles formed from it.

As illustrated below are two right triangles of the same hypotenuse of length C, the invariant is according to the equation $C^2 = A^2 + B^2 = A'^2 + B'^2$. The lengths A and B differ from A' and B', but their areas, being the sum of their lengths squared, are of the same areas as that of the hypotenuse length C squared.

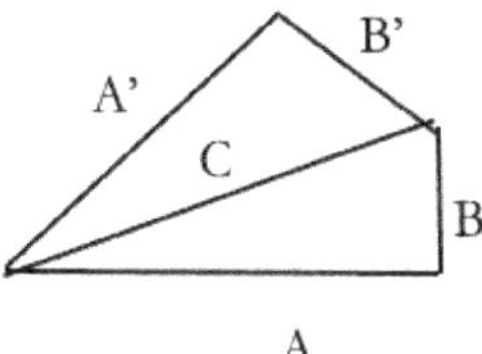

This invariance also applies to higher dimensions, as according to the rectangular box below. Length OP is the hypotenuse of a right triangle with its other sides as x and y. Length OQ is also a hypotenuse of a right triangle with its other sides as OP and z. Length OQ is thus according to the legs of two right triangles of sides x, y and z.

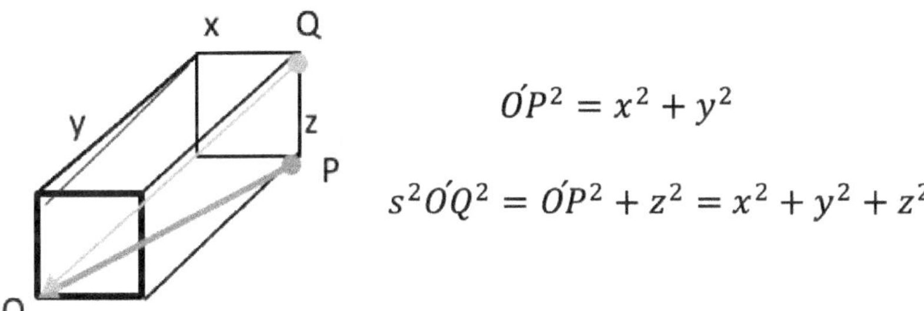

$$O'P^2 = x^2 + y^2$$

$$s^2 O'Q^2 = O'P^2 + z^2 = x^2 + y^2 + z^2$$

Physics includes, among other things, time and motion, and with SRT first formulated by Einstein and then given geometrical interpretation by Hermann Minkowski (1864–1909), time became referred to as a fourth dimension of the space-time interval s in the manner

$$s^2 = x^2 + y^2 + z^2 - c^2 t^2 = x'^2 + y'^2 + z'^2 - c^2 t'^2$$

The interval s squared equates invariance of four-dimensional space-time of two different coordinate systems. Its usefulness is similar to that of the Pythagorean theorem.

Invariance of the interval derives from the Lorentz transformations:

$$x' = \frac{x-vt}{\sqrt{1-\beta^2}} \qquad y' = y \qquad z' = z \qquad t' = \frac{t-\frac{vx}{c^2}}{\sqrt{1-\beta^2}}$$

For the derivation, time coordinates convert into distance coordinates in the manner

$$ct' = \frac{ct-\beta x}{\sqrt{1-\beta^2}}$$

The coordinates are then added and subtracted in the manner

$$ct' + x' = \frac{ct-\beta x}{\sqrt{1-\beta^2}} + \frac{x-vt}{\sqrt{1-\beta^2}} = \frac{ct-vt+x-\beta x}{\sqrt{1-\beta^2}} = \frac{ct(1-\beta)+x(1-\beta)}{\sqrt{1-\beta^2}}$$

$$ct' - x' = \frac{ct-\beta x}{\sqrt{1-\beta^2}} - \frac{x-vt}{\sqrt{1-\beta^2}} = \frac{ct+vt-x-\beta x}{\sqrt{1-\beta^2}} = \frac{ct(1+\beta)-x(1+\beta)}{\sqrt{1-\beta^2}}$$

The product (t' + x')(t' − x') gives

$$t'^2 - x'^2 = \left[\frac{ct(1-\beta)+x(1-\beta)}{\sqrt{1-\beta^2}}\right] \cdot \left[\frac{ct(1+\beta)-x(1+\beta)}{\sqrt{1-\beta^2}}\right]$$

$$\frac{c^2t^2(1-\beta^2)-ctx(1-\beta^2)+cxt(1-\beta^2)-x^2(1-\beta^2)}{1-\beta^2} = c^2t^2 - x^2$$

Hence, invariance of the interval is of the form

$$s^2 = c^2t'^2 - x'^2 = c^2t^2 - x^2$$

$$\frac{s^2}{c^2} = t'^2 - \frac{x'^2}{c^2} = t^2 - \frac{x^2}{c^2}$$

$$\frac{s^2}{t^2} = c^2 - \frac{x^2}{t^2} = c^2 - v^2$$

$$\frac{s^2}{t'^2} = c^2 - \frac{x'^2}{t'^2} = c^2 - v^2$$

Invariance of the interval is a means of relating different phenomena. In the same manner of relating spacetime coordinates, for instance, momenta P = mv and mc, and energy E = mv² and mc² are invariants according to the equations

$$\frac{s^2}{c^2} = P_x^2 + P_y^2 + P_z^2 - \frac{E^2}{c^2} = P_{x\prime}^2 + P_{y\prime}^2 + P_{z\prime}^2 - \frac{E'^2}{c^2}$$

$$s^2 = P_x^2 c^2 + P_y^2 c^2 + P_z^2 c^2 - E^2 = P_{x\prime}^2 c^2 + P_{y\prime}^2 c^2 + P_{z\prime}^2 c^2 - E'^2$$

Total energy or momentum of systems thus calculates by knowing variables of energies or momenta of one system in comparison to another.

Relativity invariance is applied according to whatever information can be obtained by it. Its purpose in this book is to understand the difference and similarity of how it relates to both special and general relativity. According to special relativity, light speed is constant relative to someone relatively at rest in gravitational free space. If s = 0, then it relates in the manner.

$$0 = c^2 t^2 - x^2$$

$$x^2 = c^2 t^2$$

$$\frac{x^2}{t^2} = c^2$$

$$\frac{x}{t} = c$$

According to general relativity, however, light speed varies according to the field strength of a gravitation field. What is now to be explained is how it relates to variable light speed regarding invariance of general relativity in analogy to both symmetry and asymmetry of covariance. The asymmetrical analogy refers to the clock paradox. Gravity is asymmetrical insofar as its relativistic effects perceived from an observer apart from it are opposite of what observers in the field of gravity perceive apart from it.

Invariance of Spacetime Curvature

The conception of Newton's inverse square law of gravity as acting at a distance is contrary to Einstein's explanation of it as spacetime curvature. Action at a distance is instantaneous, but transport of information faster than light speed is contrary to the condition of special relativity. Einstein thus generalized covariance of spacetime for it to comply with gravitational effect, as general covariance, which maintains invariance of the laws of physics in being the same for all reference frames.

A field theory refers to magnitudes in space and time differing according to temperature, gravitational effect and so forth. A convenient way of relating them is by a mathematical matrix, and the general form of Einstein's field equation is a tensor matrix for spacetime curvature as

$$G_{\mu\nu} = \frac{8\pi G}{c^4} T_{\mu\nu}$$

The left side of the equation is Einstein's tensor matrix, and the right side is a stress-energy tensor matrix in compliance with a Minkowski spacetime invariance for momentum and energy. The $8\pi G/c^4$ is an Einstein constant including Newton's constant G and light speed c. The $8\pi G/c^4$ is a more complex constant for including the interrelations of such other factors as electromagnetic radiation in terms of light speed. It has a value of about 2×10^{-48} cm/sec. It's numerical value in ratio to e^2, the electromagnetic constant's value of 23.07×10^{-20}, is nearly that of the numerical value of the electron measured according to units of measure as grams, centimeters and seconds.

However, the chosen unit of measure does determine different numerical outcomes. In relation to mass m and distance r, the ratio $G/c^4 e^2$ relates as $(v^2 r/m)/c^4(v^2 mr) = t^2/r^2$. It is thus per velocity squared or Gm/r.

The 8π relates to the addition of four circles, $4(2\pi)$, that represent four dimensional spacetime curvature. Twice r and twice m thus have one-fourth as much spacetime curvature in proportion to c^4.

The subscripts $\mu\nu$ of the matrices are similar to spacetime coordinates x, y, z, t of relative motion, but they apply more specifically to Riemannian curved space apart from Euclidian spacetime coordinates, and they are also more complex in describing positions in four dimensional spacetime instead of only coordinate directions for relative motion. The matrix thus allows for a more descriptive geometry according to the distribution of mass-energy.

Einstein's spacetime curvature is according to a Riemann curvature tensor calculus. A tensor of the first rank is a scalar, which is a magnitude of something—such as temperature or mass quantity—that can be described by a single number as a magnitude. A tensor of the second rank is a vector for including direction along with magnitude. Higher ranked tensors are more complex. The Riemann tensor, for instance, describes the shortest path along a curved surface according to a geodesic matrix $g_{\mu\nu}$.

Calculus itself is a convenient means of adding to and eliminating factors according to particular conditions. By integration, for instance, constants as other variables of influence can be added. Justification of their addition is because differentiation as opposite to integration cancels out constants being of the first order. Differentiations of x^1, x^2, x^3, for instance, become (0), $2x^1$, $3x^2$.

The spacetime curvature in general relativity includes the $g_{\mu\nu}$, as to be determined by the stress energy tensor. By substitution, the Einstein tensor on the left side of the equations becomes

$$R_{\mu\nu} - \frac{1}{2} R g_{\mu\nu} = \frac{8\pi G}{c^4} T_{\mu\nu}$$

The Riemann curvature tensor matrix $R_{\mu\nu}$ along with the scalar magnitude R and geodesic curvature tensor matrix $g_{\mu\nu}$ are thus determined according to the value of the momentum-energy tensor matrix $T_{\mu\nu}$.

$T_{\mu\nu}$ is an invariant of a stress momentum-energy tensor for describing the effects of all forms of energy. Its applications are complex, and there are various solutions as to whether the field rotates and so on.

The Schwarzschild Metric

A simpler metric was derived by Karl Schwarzschild (1873–1916). It includes the Newtonian form of gravitational escape speed replacing velocity in the relativistic factor for gravitational modification of spacetime conditions. It is of the form

$$g = \left[1 - \frac{2GM}{rc^2}\right] c^2 dt^2 - \frac{dr^2}{\left[1 - \frac{2GM}{rc^2}\right]} - r^2 d\theta^2 - (sin^2\theta)d\varphi^2$$

The last two terms of the metric with trigonometric quantities θ and φ refer to polar coordinates in place of the perpendicular coordinates y and z of flat spacetime. Flat spacetime refers to gravitational free space whereby conditions of SRT apply.

A distinction of the form of the Schwarzschild metric from that of Lorentz invariance is with regard to acceleration whereby the speed of light is derived as slower within a gravitational field. Such difference in light speed is calculable in relation to the infinitesimal increment g becoming zero by means of differentiation. It is zero in the sense infinitesimal influence of other factors in relation to gravitational free space cancel out for part of the metric to equate in the manner

$$0 = c^2 dt^2 \left[1 - \frac{2GM}{rc^2}\right] - dr^2 \left[1 - \frac{2GM}{rc^2}\right]^{-1}$$

$$dr^2 \left[1 - \frac{2GM}{rc^2}\right]^{-1} = c^2 dt^2 \left[1 - \frac{2GM}{rc^2}\right]$$

$$\frac{dr^2}{dt^2} = c^2 \left[1 - \frac{2GM}{rc^2}\right]^2$$

$$\frac{dr}{dt} = c \left[1 - \frac{2GM}{rc^2}\right] = c'$$

The relative speed c' of light in a gravitational field is thus less than speed c in relatively gravitational free space. If $2GM/r = c^2$, then c' = 0 for not even light to be able to escape from the field.

The Black Hole Singularity

Slowing of light speed in a gravitational field indicates the condition of a singularity, also known as a black hole whereby nothing, including light, can escape. It is conditional to a critical Schwarzschild radius R of the equation $2GM/R = c^2$ as an event horizon whereby nothing within it is able to escape. Since light cannot escape, it appears black.

(Regarding energy conservation with regard to light energy decreasing with an increase in gravity, consider the possibility that photon energy converts to graviton energy.)

The black hole was challenged and modified in the early 1970s by Stephen Hawking (1942–2018) for it to emit Hawking radiation. In explaining how black holes radiate, Hawking assumed a probability condition of quantum mechanics applies to allow the probability of a light particle inside a black hole to exist outside of it as well.

Although the black hole is also not refuted here, modification of theory is indicated with regard to the Schwarzschild metric indicating slower light speed. If escape speed squared equals one-half light speed squared, then light speed is slowed down to become one-half light speed.

Modification needs to maintain consistency with relative motion of Special relativity. The gravitational analogy is with regard to the ratio of escape speed squared being twice orbital speed squared as gravitational potential. If escape speed squared is $2GM/R = c/2$, then reduced light speed calculates in the manner $c' = c[1 - \frac{1}{2}] = c/2$. It squared becomes $c^2/4$. Coincidentally, gravitational potential GM/R is also one-half escape speed squared $2GM/R$, and the same as reduced light speed squared. However, local clocks and measuring rods are relatively slower and shorter whereby both rotation and light speed are reduced by the same factor. Local observers thus perceive orbital speed as $c/2$ and the reduced light speed as c.

How, then is reaching an escape speed of c prevented?

Uniqueness is according to light speed energy. If escape speed squared becomes one-half light speed squared, then slower light speed squared equals orbital speed squared as one-fourth light speed squared. The ratio of gravitational potential to slower light speed energy is thus unity. However, if escape speed squared equals one-third light speed squared, then slower light speed equals two-thirds light speed, and it squared equals four-ninth light speed squared in contrast to one-sixth of orbital speed squared. Their ratio is 8 to 3, being four-ninth divided by one-sixth. Of significance is that slower light speed energy is still greater.

To the contrary, if escape speed squared equals two-thirds light speed squared whereby c' equals one-third of c, and c'^2 is one-ninth c^2. Gravitational potential is one-half escape speed squared, being one-third of two-thirds. The ratio of one-third to one-ninth is three. There is thus three times more light energy lost than gained by the condensing of mass to obtain a higher escape speed. It is contrary to conservation of internal energy.

Special and General Relativity Analogies

The singularity itself can be in compliance with covariance whereby observers within it cannot see out of it as well as observers outside it are unable to see what is inside it. Covariance is part of the expanding universe theory whereby observers at its edge perceive themselves to be at its center. Such result occurs because of gravitational effect canceling itself out at its center. Light moving outward thus has similar gravitational influence of light moving inward.

Even though light speed varies in a gravitational field, there are symmetrical analogies of gravity and relative motion. Local light speed, for instance, can actually be perceived by observers in the field the same as in gravitational free space. Likewise, the moon that orbits around Earth represents a natural clock that is not perceived as slower because of it and Earth both nearly being slowed the same by the sun's gravitational field.

Another condition inclusive of the effect is the relativistic contraction of length. Because light speed in the field is slower by the relativistic factor squared, it moves slower also a shorter distance for it to be observed of the same frequency. Slower clocks and shorter measuring distances thus nullify the perception of slower light speed in a gravitational field.

To the contrary of observers in gravitational free space perceiving events occurring slower or of longer distance, according to interpretation of perception, are events outside the field being relatively perceived as faster or occurring at shorter distance. The difference in observable effect is analogous to the clock paradox of special relativity explained according to asymmetry. As the clock changing direction is the one determined as slow, clocks within the gravitational field are likewise determined as slow by outside observers due to the asymmetry of gravitational acceleration. For symmetry of opposite effect, as for clocks outside the gravitational field to be perceived relatively slower, they need to be within another gravitational field of greater strength.

With covariance of a singularity allowing observers to perceive themselves to be at its center, it becomes evident that observers inside it cannot see what is outside it. The singularity itself becomes a world within another world. Outside observers cannot see inside it and inside observers cannot see outside it. With evidence of black holes existing in other parts of the observable universe, worlds within worlds are thus another possible existence beyond the reality of the universe in which we live.

9

QUANTUM ORIGINS

An origin of the quantum is that of complete absorption by a black-body. Formulas were derived to determine a relation between temperature and intensity of radiation that a body of mass can absorb and emit while in equilibrium with the forces of nature. Classical formulas for predicting results of experiment failed until a concept of the quantum was introduced. Understanding this development is according to particular relations between heat and light.

Heat and Light

In the year 1800, William Herchel (1738–1822) used a variety of eye glasses of different color lenses to peer through a telescope and observe the sun. By using these different glasses of various color, he discovered light filtered through some of them felt warmer than if filtered through others. He followed up with the implications of this discovery by devising experiments consisting of prisms and thermometers to further discover a one-degree higher temperature just beyond the red end of the visible spectrum into the infrared. From further experiment of observation, he concluded invisible light energy exists beyond the visible spectrum

A heat-light connection was thus established, but it would be another twenty years before Andre-Marie Ampere (1775–1836) suggested light and heat are only different aspects of the same process. A response to Ampere's suggestion came with Marcedonio Melloni (1798–1854) agreeing inasmuch as he believed both are waves propagating through media. He regarded light as the harmonious waves of ether, and he similarly regarded heat as radiant waves of caloric, but his experimental findings from 1833 to 1840 indicated no essential difference in wave properties of these two phenomena. To his credit, he did discover refractive properties of thermal radiation.

More development followed from experiments by Jean Bernard Leon Foucault (1819–1868) and Armond Hippolyte Louis Fizeau (1819–1896) in confirming wave properties of radiant heat. Their experiments split rays of infrared light for them to superimpose and produce alternating bands of hot and cold in analogy to the light and dark fringes of ordinary light. James David Forbes (1809–1861) then discovered heat polarizes similar to that of light, and he advanced the concept of a continuous radiation spectrum that later became essential to Maxwell's theory of electromagnetism. It followed that the unification of electromagnetism and thermodynamics was in order. It would have been routine except for the predominance of wave theory at the time. The relation between vibrant molecules of matter with the waves of light was not adequately understood as of yet.

There had been some attempts to unify theory as such. Leonard Euler (1707–1783) proposed the principle a particular substance of mass can absorb light of any frequency that its smallest particle is able to vibrate. He attempted to explain phenomena according to the ether from which matter forms. His theory was not successful, but his absorbing principle did not go unnoticed.

More essential to the development of theory were discoveries of how matter absorbs light. William Wollaston (1766–1825) discovered in 1802, for instance, that light spectra emitted from matter include dark lines. Joseph Fraunhofer (1787–1826) made more discoveries along this line in 1814. The interest in these discoveries grew among theorists. Stokes, for one, used the principle of Euler to explain them as atoms absorbing light waves by means of resonance.

Pierre Prevost (1751–1839) had provided evidence indicating all bodies radiate heat. The evidence suggested further that poor absorbents of heat are also poor emitters of it and good absorbents are good emitters, and it became evident materials in thermal equilibrium emit what they absorb. In 1858, Balfour Stewart stated the law that the absorptivity of a material in a state of thermal equilibrium is equal to its emissivity. Stewart assumed the absorptive and emissive ability of different materials varies in relation to the nature of their internal substance. Gustav Kirchhoff (1824–1887) proposed in 1859 a particular condition of "black-body" radiation applies to all bodies regardless of their material composition.

Kirchhoff examined the spectrum of sunlight through a sodium flame to discover dark lines of the spectrum change to yellow when the sunlight is of low intensity, being darker with more intense sunlight. He also found sodium emits the same part of the light spectrum absorbed with an appropriate increase in temperature of the sodium flame. Kirchhoff thus surmised the ability of substance to absorb and emit a certain color of light depends on its relative state of equilibrium. Further experiments to confirm this premise indicated absorption and emissivity for a material is a function of its temperature and wavelength or frequency.

The mathematical formulation of the law assumes a system obtains the state of thermal equilibrium at sub temperature below incandescence. Thus, if A denotes the total radiation per surface area on each body of mass, a the fractional amount of radiation actually absorbed by its material, and if E is the permissible radiation emitted from it, then the relations are mathematically expressed in the manner

$$aA = E \qquad A = \frac{E}{a}$$

For all materials in a state of thermal equilibrium, the powers of emissivity (E_1, E_2, E_3, etc.) divided by their respective factors of absorption (a_1, a_2, a_3, etc.) equal the same amount of light incident per surface area:

$$A = \frac{E_1}{a_1} = \frac{E_2}{a_2} = \frac{E_3}{a_3} \cdots = \frac{E_n}{a_n}$$

Kirchhoff defined the black-body as one absorbing and emitting radiation of all frequencies or wavelengths, but the power of emissivity of the black-body is E_B, and the fraction of light absorbed is unity, $a_B = 1$, whereby

$$\frac{E}{a} = \frac{E_B}{1}$$

Dividing by E_B, multiplying by a, changing order and assuming absorption equals emissivity relates in the manner

$$a = \frac{E}{E_B} = \varepsilon$$

The symbols a and ϵ denote absorption and emissivity, respectively, of a material body in ratio to a black-body.

Further relations are of the discovery of a fourth power law commonly referred to as the Stefan-Boltzmann Law.

The Stefan-Boltzmann Law

The rate systems change from one temperature to another to obtain a state of equilibrium with its environment is further significant. Newton had assumed the process is linear inasmuch as the rate of change in temperature is proportional to the difference in temperatures between the system and its environment, but experimental data indicated the relation is approximate, only true at relatively normal temperatures, as the data did not appear linear at higher ones.

Another relation superseded Newton's in the late nineteenth century. It came from a study of temperature and light intensity by John Tyndall (1820–1893) and Joseph Stefan (1835–1893). Tyndall had run an electric current through a platinum wire that resulted in heating the wire to a state of incandescence. By measuring the radiation emitted at different temperatures, he found the light intensity to be about twelve times greater with a wire being about 1200 degrees centigrade than if only 525 degrees centigrade. He calculated

$$\left[\frac{273°+1200°}{273°+525°}\right]^4 = \left[\frac{1473°}{798°}\right]^4 \approx 12$$

The $273°$ is the centigrade scale above absolute zero. The ratios thus relate to absolute zero.

This mathematical relation appeared the same for all substances at all temperatures. Stefan concluded the intensity I is proportional to the fourth power of the absolute temperature T:

$$I = k_B T^4$$

Constant k_B is named the Boltzmann constant. Its value is 1.3806503 X 10^{-23} J/K, which is 1.38065 X 10^{-16} grams multiplied by centimeters squared per seconds squared per one-degree Kelvin.

Because of such minuteness of the numerical value of the constant, a slight change in temperature, even to the fourth power, causes only a slight change in light intensity. However, Tyndall's measurements were not accurate. Later results gave a ratio of about 18 to 1, but not according to Kirchhoff's black-body condition by which Ludwig Boltzmann deduced the law in accordance with the second law of thermodynamics named entropy.

Boltzmann's derivation includes an exchange of heat between different systems. System S_2 receives heat energy Q_2 from system S_1. The loss of Q_2 in ratio to absolute temperature T_2, as negative entropy, equals the amount of positive entropy Q_1/T_1 remaining in S_1. Along with the exchange of heat are differences in size of energy densities. Energy density u_1 of S_1 is reduced in size from V_1 to V_4, and energy density u_2 is increased in sized from V_2 to V_3.

A change in state of heat energy can be interpreted in terms of work performed. Work energy by means of an adiabatic process further relates to conservation of entropy. The process can further be in compliance with Boyle's law whereby a product of pressure p and volume V of a gas at given temperature T is constant: $pV = k$. Pressure further relates in terms of force F, such as centripetal force of gravity maintaining pressure p inside a tire per surface area $4\pi r^2$. Surface area in ratio to volume equates in the manner $4\pi r^2/(4\pi r^3/3) = (1/3)r$. However, comparing surface area to other surface area and volume to other volume is, respectively, just V_1 per V_2. Regarding the latter condition, the ratio of gravitational force, in relation to potential energy mv^2 within volume V, relates to the constant k of pressure-volume pV in the manner

$$pV = \frac{FV}{r^2} = Fr = FV^{\frac{1}{3}} = k$$

Energy density u of particles within the volume, of potential mass-energy mv^2, relates per volume in the manner

$$u = \frac{F}{V} \qquad F = uV$$

By substituting uV for F in the previous equations, it becomes

$$uVV^{\frac{1}{3}} = uV^{\frac{4}{3}} = k$$

The square root of the square root of the equation cubed relates it in terms of volume space as

$$u^{\frac{3}{4}}V = k^{\frac{3}{4}}$$

Entropy is next considered in relation to different temperatures. States S_1 and S_2 relate as conserved entropy in the manner

$$\frac{Q_1}{T_1} + \frac{Q_2}{T_2} = \Delta S$$

However, in consideration of the amount of change occurring of each system, it is just one amount passed onto the other. Hence

$$\frac{Q_1}{T_1} - \frac{Q_2}{T_2} = 0$$

The difference in entropy states Q_1 and Q_2 further relate to different energy densities and change in size in the manner

$$Q_2 = u_2(V_3 - V_2)$$

$$Q_1 = u_1(V_1 - V_4)$$

Hence

$$\frac{Q_2}{T_2} + \frac{Q_1}{T_1} = \frac{u_2(V_3 - V_2)}{T_2} + \frac{u_1(V_1 - V_4)}{T_1} = 0$$

In relation to Boyle's Law, which requires different constants for different temperatures and sizes of energy densities, k_1 and k_2 become the respective constants in the manner

$$u_2^{\frac{3}{4}}V_3 - u_2^{\frac{3}{4}}V_2 = k_2^{\frac{3}{4}} - k_1^{\frac{3}{4}} \qquad u_1^{\frac{3}{4}}V_1 - u_1^{\frac{3}{4}}V_4 = k_1^{\frac{3}{4}} - k_2^{\frac{3}{4}}$$

$$u_2^{\frac{3}{4}}(V_3 - V_2) = k_2^{\frac{3}{4}} - k_1^{\frac{3}{4}} \qquad u_1^{\frac{3}{4}}(V_1 - V_4) = k_1^{\frac{3}{4}} - k_2^{\frac{3}{4}}$$

$$V_3 - V_2 = \frac{k_2^{\frac{3}{4}} - k_1^{\frac{3}{4}}}{u_2^{\frac{3}{4}}} \qquad V_1 - V_4 = \frac{k_1^{\frac{3}{4}} - k_2^{\frac{3}{4}}}{u_1^{\frac{3}{4}}}$$

Substituting the right-hand sides of these latter equations into the thermodynamic equation obtains

$$\frac{Q_2}{T_2} - \frac{Q_1}{T_1} = \frac{u_2(V_3-V_2)}{T_2} - \frac{u_1(V_1-V_4)}{T_1}$$

$$\frac{u_2}{T_2}\left[\frac{k_2^{\frac{3}{4}}-k_1^{\frac{3}{4}}}{u_2^{\frac{3}{4}}}\right] - \frac{u_1}{T_1}\left[\frac{k_1^{\frac{3}{4}}-k_2^{\frac{3}{4}}}{u_1^{\frac{3}{4}}}\right]$$

$$\frac{u_2^{\frac{1}{4}}}{T_2}\left(k_2^{\frac{3}{4}} - k_1^{\frac{3}{4}}\right) - \frac{u_1^{\frac{1}{4}}}{T_1}\left(k_1^{\frac{3}{4}} - k_2^{\frac{3}{4}}\right)$$

$$\frac{u_2^{\frac{1}{4}}}{T_2}\left(k_2^{\frac{3}{4}} - k_1^{\frac{3}{4}}\right) - \frac{u_1^{\frac{1}{4}}}{T_1}\left(k_2^{\frac{3}{4}} - k_1^{\frac{3}{4}}\right) = 0$$

Hence

$$\frac{u_2^{\frac{1}{4}}}{T_2} - \frac{u_1^{\frac{1}{4}}}{T_1} = 0 \qquad \frac{u_2^{\frac{1}{4}}}{T_2} = \frac{u_1^{\frac{1}{4}}}{T_1}$$

Because energy densities and temperatures are equal, the adiabatic change is constant, such that

$$\frac{u^{\frac{1}{4}}}{T} = k^{\frac{1}{4}} \qquad u = kT^4$$

Moreover, u is energy density of the radiation of intensity I absorbed or emitted as either proportional to a change in energy per volume or the intensity of a radiation incident on a black-body. They express different aspects of energy, but they are of the same process. By substituting I for u, $I = k_B T^4$ interprets as the Stefan-Boltzmann's law, with k_B designated as the Boltzmann constant.

Wien's Displacement Law

The Stefan-Boltzmann law relates the difference in intensity of radiation to temperature, but it does not specify how either frequency or wavelength of light applies. Wilhelm Wien (1864–1928) determined the relation for the intensity of radiation for a temperature and the range in the wavelength or frequency of the temperature. He noted shorter wavelengths are more frequent, numerous and energetic than longer waves.

For further analyses, Wien adopted the adiabatic concept in relation to black-body radiation. However, a true black-body is not available to study within the confines of the laboratory. Although soot is black, for instance, it still emits a radiation invisible to the eye. Even so, black-body conditions of equilibrium tend to occur with a slight change in the environment from its interaction with radiation. Although an oven allows radiation to escape, the oven temperature can still be maintained by means of using fuel. Wien thus considered a hollow container staying in thermal equilibrium by it having a tiny hole at its surface to allow radiation to enter and reflect here and there on the walls of the container before it finally escapes.

Wien surmised a black-body varies with a small range in wavelength of radiation at maximum intensity in inverse proportion to temperature. A shorter more frequent wave at maximum intensity thus correlates with a higher temperature in the manner $\lambda_{max}T = b$ as Wien's displacement law, where b is a constant.

Relating the intensity of radiation along with both the temperature and wavelength is more complex. Although total intensity is to the fourth power of temperature, determining it according to any particular wavelength at any given temperature needs to include three variables as length, intensity and temperature. Since total intensity is to the fourth power of temperature, since the wavelength shortens per higher temperature, and because of such other considerations, as with regard to the Doppler effect, Wien eventually concluded the intensity for a particular wavelength at a given temperature is to the fifth power of the wavelength.

To derive a formula, he related wavelength and temperature according to a distribution law derived by Maxwell for relating molecular speeds of gas particles in relation to temperature:

$$dN = \left[\frac{m}{2\pi T}\right]^{\frac{1}{2}} \cdot e^{\frac{-1}{2}mv^2/k_B T}$$

The exponential function e ≈ 2.71828182 in Maxwell's formula is raised to a power of kinetic energy per $k_B T$, a product of the Boltzmann constant k_B and temperature T.

Wien associated cavity radiation as molecular resonance and frequency in relation to kinetic energy to derive a function F(λ,T) for temperature and range in wavelength λ with the insertion of constants a and b in relation to the intensity I_λ:

$$I_\lambda \cdot d\lambda = \frac{a}{\lambda^5} \cdot e^{-b/\lambda k_B T}$$

It fared well with data in relation to higher frequencies but not in relation to lower ones.

Planck's Solution

Scientists had been aware from the mid-nineteenth century that light escaping from an oven of higher temperature through a tiny crack is more energetic. Therefore, Ferdinand Kurlbaum (1857–1927) and Heinrich Leopold Rubens (1865–1922) experimented in observing waves as long as 59 microns, which is one twentieth of a millimeter. Wien's formula failed in predicting accurate results for these longer wavelengths.

When Rubens revealed the experimental results to Max Planck (1858–1947) in 1900, Planck had a solution the same day. He assumed resonators of radiant heat mediate between molecules and radiation for absorbing, storing and releasing the same particular quantity of radiant energy. Planck's solution to the problem was therefore to quantize energy as multiples of hf, wherefore h is now known as Planck's constant, and f is frequency in relation to energy.

The amount of energy of an oscillator is according to the values of each energy level, which are 0, hf, $2hf$, etc. With some values as zero, neither the energy level nor the spatial distribution is continuous. The distribution itself is determined according to an exponential function of the Boltzmann factor

$$e^{\frac{-E}{k_B T}}$$

As to further develop Maxwell's statistical treatment of the kinetic theory of gases, Boltzmann had derived the probabilities in terms of the exponential function, and he also applied discrete energy levels as infinitesimal divisions among the actions of molecules. Planck instead assumed non-infinitesimal-discrete-energy-levels apply.

A number N of oscillators in an incremental range of frequency near f has various multiples of energy E = hf, including zero, and is the number n times each consecutive level of the probability distribution in accordance with

$$N = n + ne^{-hf/k_BT} + ne^{-2hf/k_BT} + \ldots$$

$$n\left(1 + e^{-hf/k_BT} + ne^{-2hf/k_BT} + \ldots\right)$$

The infinite series in the equation converges to

$$V = n\left(e^{-hf/k_BT}\right)^{-1}$$

The total energy of all oscillators is

$$E = n(0) + n(hf)^{-hf/k_BT} + \ldots$$

$$nhfe^{-hf/k_BT}\left[1 + 2\left(e^{-hf/k_BT}\right) + 3\left(e^{-hf/k_BT}\right)^2 + \ldots\right]$$

This infinite series converges to

$$E = nhfe^{-hf/k_BT}\left(1 - e^{-hf/k_BT}\right)^{-2}$$

Dividing E by N obtains

$$\frac{E}{N} = \frac{nhfe^{-hf/k_BT}\left(1-e^{-hf/k_BT}\right)^{-2}}{n\left(1-e^{-hf/k_BT}\right)^{-1}} = \frac{hfe^{-hf/k_BT}}{1-e^{-hf/k_BT}} = \frac{hf}{e^{-hf/k_BT}-1}$$

E/N represents the average energy per oscillator of incremental frequency range between f and $(f + df)$ inasmuch as n factors out of the equation.

Comparing Formulas

In comparison to Wein's formula, Planck derived

$$I_\lambda \cdot d\lambda = \frac{2c}{\lambda^5} \cdot \frac{h}{e^{hc/\lambda k_BT}-1}$$

A main difference in it and Wein's formula is Wein's constants *a* and b are replaced with 2hc and a different form of the exponential function. As for 2hc, it is twice a change in speed c with regard to a reverse in direction.

The constant h relates dimensionally as a product of mass, velocity and distance, whereas the constant k_B is the product mass and velocity squared. Thus, hc is dimensionally the same as λk_B, such that they are of the same ratio no matter what are units of mass, distance and time used to determine results. The exponential function thus changes only by the temperature ratio of T, which is consistent with Wien's displacement law. If $c = \lambda f$, then replacing each λ with c/f obtains

$$I_f \cdot df = \frac{2f^5}{c^4} \cdot \frac{h}{e^{hf/k_B T} - 1}$$

Another formula was derived as

$$I_f \cdot df = \frac{2f^3}{c^3} \cdot \frac{h}{e^{hf/k_B T} - 1}$$

However, these two formulas differ by the ratio $(2f^5/c^4)/(2f^3/c^3) = f^2/c$.

These formulas pertain to different forms of energy. Planck's form is a modification of Wien's formula and the other one is from a formula derived by Sir James Jean (1877–1946) and John William Strutt (1842–1919), renamed Lord Rayleigh. Wien's distribution formula is in accordance with his displacement law for a change in temperature; the other formula was derived in accordance with an equal partition theorem that had been derived independently first by John James Waterston (1811–1883) and later by Maxwell and Clausius for advancing the kinetic theory of gases. It relates to degrees of freedom.

A degree of freedom relates in accordance with the number of modes of vibrations. The same space is assumed to be more capable of containing a greater number of modes of shorter wavelengths. Rayleigh assumed there is a tendency for shorter modes to dominate. However, the shorter waves are more energetic because of their shortness allowing for a greater number of more rapid vibrations, such as for it to result in an ultraviolet catastrophe from a tendency to increase to infinite energy. Such a result was considered in violation of conservation of energy, and the formula differs from the Stefan-Boltzmann fourth power law. The formula was thus in need of modification.

In Planck's formula, change in wavelength λ relates in the manner

$$\frac{hc}{\lambda^5} = \frac{hf\lambda}{\lambda^5} = \frac{hf}{\lambda^4} = \frac{mc^2}{\lambda^4}$$

The change in radiant energy is thus per fourth power of wavelength. In the other formula, change in frequency f relates in the manner

$$\frac{hf^3}{c^3} = \frac{mc^2 f^2}{c^3} = \frac{mc^2 f^2}{c(\lambda^2 f^2)} = \frac{mc}{\lambda^2}$$

There is thus a change in momentum occurring per wavelength squared in contrast to internal energy of mass per wavelength of the fourth power. One simply relates more directly to energy while the other one relates more directly to momentum.

10

QUANTUM PARTICLE PHYSICS

An early application of the quantum was Bohr's theory of the atom. Discoveries of photoelectric and Compton effects followed. Further advance came with the development of quantum wave mechanics whereby wave equations became interpreted as probability equations according to an uncertainty principle. Such new interpretation is now established according to standard physics theory, but with such various challenges as string theory proposing extra dimensions for unified theory. Although such development is complex, quantum theory itself can still be explained simpler according to the theory of the Bohr atom in establishing the fundamental relations of physical constants.

The Bohr Atom

Applying quanta to the structural nature of the atom came from Niehs Henrik David Bohr (1885–1962). He modified the atomic model previously announced by Ernst Rutherford (1871–1937) in 1911, who had contrived a model of the atom to describe how alpha particles reflect. Accordingly, the bulk of mass that scatters is contained within a nuclear radius of about 1,836 times smaller than the radius of the atom itself. Rutherford further assumed atoms continually absorb and emit radiation as electrons circle the nucleus. However, his model failed to predict results of all phenomena.

One such phenomenon pertains to the spectra of radiation emitted by atoms. The spectral lines of light associated with atoms did not conform to a pattern consistent with a known theory of continuous spectra. Formulas had been provided apart from theory, as in 1885 by Johanne Jakob Balmer (1825–1890), and by later experiments by Johannes Robert Rydberg (1854–1919), Carl Runge (1856–1927) and Henrich Kayser (1853–1910). Their *ad hoc* formulas, for the most part, agreed with observation, but they lacked the theoretical foundation until Bohr proposed, in 1913, a quantum restriction for his modified version of Rutherford's atomic model.

Bohr assumed electrons orbit the nucleus of an atom in circular paths whereby atoms either emit or absorb radiation only when an electronic state of angular momentum changes in the quantum condition of multiple units.

Bohr only modified classical formula by applying quantum restrictions in assuming a force field exists, consistent with Coulomb's inverse square law for electrostatics and magnetism. Electrons thus have a negative unit of charge –e and nuclei have a positive unit of charge +e. An electrostatic force between them is the product of charge per distance r squared:

$$F = \frac{e}{r} \cdot \frac{-e}{r} = \frac{-e^2}{r^2}$$

In comparison, Newton's second law of motion for centripetal force relates in the manner

$$F = \frac{mv^2}{r}$$

Equating the values electrostatic charge and centripetal force as both being either negative or positive is of the manner

$$\frac{-e^2}{2r^2} = \frac{\frac{-1}{2}mv^2}{r}$$

$$K = \frac{e^2}{2r} = \frac{1}{2}mv^2$$

It thus relates to classical formula for kinetic energy K of the electron.

The potential energy E of the electron within a conservative field of force is

$$E = \frac{-e^2}{r}$$

The total energy W is therefore

$$W = K + E = \frac{e^2}{2r} - \frac{e^2}{r} = \frac{-e^2}{2r} = \frac{-1}{2}mv^2$$

$$r = \frac{e^2}{mv^2}$$

The orbital radius of the electron thus equates as the unit of charge squared per twice the kinetic energy of the electron as its potential energy.

Bohr assumed total energy W becomes zero as r becomes infinite or whenever the atom becomes ionized because of it losing an electron. By applying quantum restrictions to the atomic radius in using the relations $n\hbar = nh/2\pi = n(mvr)$ and $e^2 = mv^2r$, he deduced

$$r = \frac{e^2}{mv^2} = \frac{e^2 n^2 \hbar^2}{m^3 v^4 r^2} = \frac{e^2 n^2 \hbar^2}{me^4} = \frac{n^2 \hbar^2}{me^2}$$

The letters m represents the mass of the electron, $\hbar$ the Planck constant h divided by 2π in relation to radius instead of perimeter, and n the numerical unit one of quantum change in energy due to ionization.

The respective velocity v squared of the electron around the nucleus is

$$v^2 = \frac{e^2}{mr} = \left[\frac{e^2}{m}\right] \cdot \left[\frac{me^2}{n^2 \hbar^2}\right] = \left[\frac{e^2}{n\hbar}\right]^2$$

$$v = \frac{e^2}{n\hbar}$$

The ratio of v to light speed c derives in the manner

$$\frac{v}{c} = \frac{e^2}{n\hbar c} = \frac{\delta}{n}$$

$$\delta = \frac{v}{c}n = \frac{ne^2}{n\hbar c} = \frac{e^2}{\hbar c} = \frac{mv^2 r}{mvrc} = \frac{v}{c}$$

The fraction v/c is referred to as the fine structure constant. Its numerical value has been determined to be 1/137.036.

The fine structure constant was originally proposed in 1916 by Arnold Sommerfeld (1868–1951). Bohr's formulation only considered the orbital motion of the electron around the nucleus of the atom as circular. Sommerfeld considered it as elliptical. An ellipse indicates energy other than just that of light is possibly involved in the determination of orbital speed v.

The elliptical orbit indicates v is an average orbital speed, but further interpretation of its value is inclusive. Arthur Holly Compton (1892–1962), for instance, related the internal mass-energy of the electron to a shorter orbital frequency in relation to a shorter orbital distance of the electron: $2\pi r_a/137.036 \approx 2.42$ X 10^{-10} centimeters. Light speed c divided by this length equates as a frequency of orbit with an approximate value of 1.22 X 10^{20} times per second. This frequency multiplied by the Planck constant h = $2\pi\hbar$ = 6.626 X 10^{-27} (gm)(cm)^2/sec equates as internal mass-energy of the electron: $m_e c^2$ = (9.11 X 10^{-28} gm) X (3 X 10^{10} cm/sec)^2 = (8.199 X 10^{-7})gm(cm/sec)^2.

Further relation is according to the square of the electromagnetic constant e equaling $m_e v^2 r_a$ and $m_p v^2 r_n$ in comparison to the reduced Planck constant equaling $m_e v r_a$ and $m_p v r_n$. With v being itself constant, the products of mass and radius are also constants: $m_e r_a = m_p r_n$. The product of the electro mass and the radius of the hydrogen atom is thus the same as the product of the nuclear proton and its radius. The difference between electromagnetic charge and its quantum energy is that the former is v greater than the latter. Moreover, $e^2/\hbar c$ = v/c = $1/137.036$, being the fine structure constant.

Bohr further recalculated the energy to eliminate r from the equations. Previous equations gave

$$E = -\frac{1}{2}mv^2 \qquad v^2 = \frac{e^4}{n^2\hbar^2}$$

The highest energy level is assumed to be in relation to n = 1.

Bohr's next assumption was the atom is able to obtain a lesser energy level by emitting a photon of energy E – E' as a difference of energy levels. This energy, for a general formula predicting light spectra emitted from atoms, relates to the frequency f of the photon by the equation

$$E - E' = \hbar f = \frac{1}{2}\delta(mc^2)\left[\frac{1}{n^2} - \frac{1}{n'^2}\right]$$

A stipulation applies here for n = 1, and for n' to only be any integer greater than one, as for an atom to either decrease or increase from an energy level to another energy level by either absorbing or emitting a photon or electron as a quantum of energy.

The Photoelectric Effect

Planck interpreted the nature of the quantum as a molecular oscillator, but Einstein interpreted it as applying to light as well as to the oscillators in explaining the photoelectric effect that was discovered in 1902 by Phillepps Lenard (1862–1947). Lenard discovered the result of electrons emitted from a metal due to light shining on the metal depends more on the frequency of light instead of its intensity. No electrons are emitted if the frequency is too low.

Einstein interpreted Lenard's findings according to Planck's quantum condition of energy as discrete multiple units of hf. Light quanta, named photons, therefore collide with a metal such that the energy of each photon transfers to an electron for the electron to break loose its bondage with the metal.

Even though the electron only absorbs a particular light quantum, the possible energy of light is still continuous along with energy of matter with regard to relative motion. The continuance of light energy is associated with varying wavelength λ and varying frequency f with regard to how they interact with matter according to the Doppler principle of relative motion. The quantum relation hf increases by a relativistic factor for relative motion of matter, for instance, for allowing continuous energy of light to correspond to continuous energy of matter in relative motion such that more intense light reflected by matter can increase thermal temperature and kinetic energy, but the ejection of electrons is still determined by the quantum states of matter and light. The quantum frequency of light relates to the quantum energy needed to free electrons from their atomic bondage, even though frequency itself need only be too quick for the equilibrium state of the atom to maintain without enough time to respond.

After determining a numerical value for the fundamental charge of the electron, Robert Andrews Millikan (1868–1953) was able to verify Einstein's explanation of the photoelectric effect as well. His experiment was that of a statistical nature in measuring the electrical voltage of electron emissions in comparison to the light intensity on a metal screen. It produced a photon to electron emission in accordance with the light energy needed of a particular frequency. In 1915, Millikan's controlled experiments were able to interpret the data in a way it not only convincingly verified the photoelectric effect but also confirmed the value of Planck's constant as well.

The Compton Effect

Similar to the photoelectric effect is the Compton effect that includes the re-direction and loss of photon energy as well as an ejection of electrons from atoms. However, only reflection between light and free electrons apart from containment, such as by a metal, is observed. In explaining statistical results of electron and light interaction, Compton assumed photons and electrons are particle-like. Laws of conservation of energy and momentum thus apply to statistical results for equating with a loss of energy of photons according to their angle of deflection, and there being a gain in energy of the electrons recoiling in opposite directions.

In 1922, Compton bombarded graphite with x-rays in examining how x-rays scatter electrons. He discovered their angle of scattering is consistent with conservation of momentum. The momentum of a scattered electron at a particular angle coincides with a loss in momentum of the scattering x-ray and its recoil angle, according to a Doppler shift to a longer wavelength. In manner of conservation of momentum, a change in the wavelength of an x-ray after it collides with an electron is

$$\Delta\lambda = \lambda' - \lambda = \frac{h}{m_e c}(1 - cos\theta)$$

With straight ahead collision, the cosine angle is unity such that $1 - 1 = 0$, as zero change in wavelength. Since the right-angle reflection is a zero-degree cosine angle, the Compton wavelength for the electron is

$$\lambda_C = \frac{h}{m_e c} = 2.43 \times 10^{-10} centimeters$$

For a 180-degree reflection, the cosine angle is minus unity is a change in wavelength twice the Compton wavelength.

The Compton wavelength λ and frequency f represent properties of a photon moving at speed c. For interacting with an atomic particle, they also relate to a quantum h of energy and mass m relatively at rest in the manner

$$E = hf = \frac{hc}{\lambda} = mc^2$$

Note: Potential or twice kinetic energy K of the electron around the nucleus of the hydrogen atom is $2K = m_e v^2 = hv/2\pi r_a = \hbar v/r_a$. The fraction 2K/E is such that its square root equal to v/c further equals the fine structure constant as 1/137.036. There is thus an equilibrium state for the accumulation of material particles of less speed than that of the equilibrium state of light. The cause could be the formation of the quantum state results from the accumulation of light with a particular frequency. It determines an equilibrium state of the quantum nature of mass whereas other frequencies of light are according to continual difference in effect regarding relative motion and gravity.

Probability Interpretation

A quantum-wave theory had also been developed by Erwin Schrodinger (1887–1961). Leading physicists as Max Born (1882–1970), Werner Karl Heisenberg (1901–1976) and Niehs Bohr reinterpreted Schrodinger's wave equations as probability equations. Born proposed the wave packet associated with the intensity of the action is the probable whereabouts of a particle. Heisenberg and Bohr then associated both waves and particles according to probability.

Born reinterpreted Schrodinger's theory being a cloud of negative charge in place of Bohr's original model of the particle circling the nucleus of the atom. For instance, in the hydrogen atom, which consists of one electron as one unit of charge, the total charge of the electron cloud—as Schrodinger had envisioned—is charge -e. In addition, he proposed the product of charge and an equilibrium state of wave intensities, as the product of amplitudes, equals its energy density at any point dxdydz when the energy of the atom is in a stable state.

The stable equilibrium state allows for both positive and negative densities, but a negative density—as interpreted at the time—was not considered comprehensible as part of the natural world.

Heisenberg and Bohr interpreted the condition according to principles of uncertainty and complementary. They found it difficult in view of the wave-particle paradox to conceive of a single particle obeying interference of two wave patterns after they pass through only one of two holes. They preferred instead to view the particle as a sort of diffused cloud capable of interfering with itself or as consolidating into a particle-like form in that the nature of wave-particle effect is neither wave nor corpuscular in an ordinary sense; it manifests instead as certain observable effects of whatever it is that causes them to appear as such. A diffused cloud represents the effect when less observed; a tinier, more precise wave packet represents the effect when observed more accurately.

Although Schrodinger depicted the diffused cloud of negative charge as well, the depiction by Heisenberg and Bohr differed according to their view of an electron or any other particle inasmuch as they interpreted wave equations as probability equations of how particle effects can be observed.

An explanation of the probability condition could be it is because total energy moving through space is greater than what is actually observed. The observable secondary effects of nature, as Gassendi had proposed centuries earlier, could arise from a primary unobserved source underlying nature. However, without means of verification, the explanation is not science; it is speculation. Nonetheless, by the probability conditions of quantum physics suggesting an existence of a virtual field of virtual particles, the philosophy of Gassendi is worthy of consideration for a more complete understanding regarding wave-particle duality and physics in general.

Uncertainty

Heisenberg established a principle of uncertainty in order to determine the probability of finding the location or time of a particle effect from the probable outcome of its momentum or energy, respectively. According to Born's interpretation of Schrodinger's equations, a wave-packet defines the region where a particle can be located. The probability of the particle's position is at any point within the wave-packet where it is proportional to the product of the total volume of the wave packet and its relative intensity. The relative size of the wave packet is dependent on such observation as the ability of a photon to penetrate a certain level within the wave packet. Photons having higher frequency, greater momentum and more energy penetrate deeper into the wave packet for a more precise location of the particle. Photons with less frequency determine instead more exact momentum from the packet's larger size but can pass through a more general state of equilibrium for not enough observable effect to be aware of.

Heisenberg's uncertainty principle follows from how Born viewed the wave packet. By Born, any one of the waves in the packet is representative of a probable particle of particular momentum and energy according to the range in energy by which waves vary. The causes of their spreading apart by impact of the photon and by the difference in energies of waves render the momentum and energy more uncertain. In following this lead, Heisenberg surmised more exact determination of the particle's position by a more energetic photon causes more uncertainty of the particle's momentum and energy. Conversely, the determination of momentum and energy of a more energetic photon results in more uncertainty of position.

The relative momentum of the photon in relation to wave parameters and Planck's constant h is provided in relation to the equation $P = \hbar/\lambda$. The uncertainty of the change in momentum ΔP with regard to the wavelength of the photon is opposite to the uncertainty of its change in position Δx. Certainty of position is uncertainty of momentum; conversely, certainty of momentum is uncertainty of position. Therefore, a range of positions for each momentum and a range in momenta for each position exist. The total uncertainty is the product:

$$(\Delta x)(\Delta P) \geq \lambda(\hbar/2\,\lambda) = \hbar/2$$

Total uncertainty further includes energy and time. The dimensions of h (as mass-velocity-distance or mass-distance squared per time) can also be interpreted as the product of distance and momentum (mass multiplied by distance and velocity) or as the product of time and energy (time multiplied by mass and velocity squared):

$$(\Delta t)(E) \geq \hbar/2$$

This probable uncertainty is not interpreted the same as that of flipping a coin. The coin can come up heads 3 times and tails 7 times after 10 tosses. After a million flips the heads-tails ratio is more apt to approximate as 1. In contrast, the probability of quantum physics is an exact prediction. If there is a probability of a particle showing up 3 times out of 10, then it is predicted as such. It can show up one time for the first eight observations, one time for the ninth observation, and one time for the tenth observation, or otherwise, but it is predicted to show three times in any combination of ten observations.

Renormalization

Plank had derived the Plank constant as a resolution to an infinity result by means of quantization. However, another infinity paradox reappeared in view of the Heisenberg uncertainty principle along with the mass-energy equation and further discoveries of particle interaction relating to the nucleus of the atom. They led to such advanced technology as the invention of the atomic bomb. A resolution of the latter paradox is referred to as renormalization. It distinguishes between normal reality of direct observation and a virtual field of energy beyond that of direct observation.

The virtual field includes quarks interacting within the nucleus of the atom whereby emerging effects appear to occur, and there is now both positive and negative charge associated with matter and antimatter in further association with the concept of spin distinguishing photons and gravitons from particles of matter. Quarks associate with perturbation theory regarding internal action between subatomic particles within the atomic structure of matter instead of mere action of matter on other matter according to classical mechanics.

Note: mathematical infinity is undefined but does occur in mathematical formulation of theory. Consider conservation of momentum, for instance, whereby momentum is conserved of collisions between masses. If one mass is relatively at rest, then its inverse ratio to the other is the undefined infinity, and their product is an undefined zero as well. To the contrary, the resulting momentum is calculable. Consider twice mass of the other moves at unit speed. The total momentum before the collision is two momenta units. For it to be conserved of inelastic collision the total three mass units need to become two momenta units. It simply calculates as three mass units multiplied by two-thirds speed units. The momenta ratios before and after inelastic collision are the same even though momenta ratio or product before collision are infinity and zero. Renormalization is thus justified.

Quantum Singularity Implications

A singularity is indicated of the Schwarzschild metric of general relativity in that on one side of the equation the increment dt equaling zero directly relates to dr equaling infinity. Special relativity, however, is conditional to light speed c being a limiting factor that matter can neither equal nor succeed. To increase to infinite mass-energy at light speed, the mass needs to absorb infinite mass-energy from the mass or electromagnetic energy accelerating it. However, perturbation occurs according to quantum physics in the manner of virtual particles having change in effect according to mathematical probability, and particles having greater mass-energy are of smaller size. Virtual energy thus approaches infinity for the wavelength of a virtual photon approaching zero. The potential of greater energy being smaller in size for less detection of it in the same amount of surrounding space allows for a singularity to exist as a particle except for the probability of detecting infinite energy of a photon wave of zero length being itself zero. However, virtual energy not having directly observable effect allows its true origin to be beyond the equilibrium state of the natural world in which we live.

The quantum condition is according to two particular speeds, one of light and the other of the quantum energy relating to a speed approximately 137.036 times less than light speed. Regarding relative motion and gravity, they are not analogically the same. Because light speed decreases within a gravitational field whereby it can even become zero, it is thereby interpreted as a black hole whereby no other mass-energy other than gravity itself is able to escape the gravitational field.

The gravitational field is also a measure of entropy. Hawking thereby challenged the black hole as contrary to possible entropy conditions, and there is a possible solution for entropy to be conserved. Part of the solution is that an increased measure of it in one system of equilibrium need only be compensated by a decreased measure of it in another system of equilibrium. Another part of it is that it requires a recycling process. For a vacuum effect of gravity to occur, there needs to be a positive force pushing inward. Two different states of equilibrium thus combine as an equilibrium state whereby relatively massless gravitons are continually created to gradually convert back to the positive-repulsive force.

Renormalization of virtual energy can similarly be consistent with special relativity regarding entropy. The infinite possibilities of infinite-positive-fields of energy could be nullified by infinite possibilities of infinite negative fields, as can infinite mass-energy of matter be nullified by infinite mass-energy of antimatter being further counterbalanced by an infinite volume of space in which to interact.

What has developed of quantum electrodynamics (QED) from these new discoveries is quantum chromodynamics (QCD). It is a theory of strong interaction between quarks and gluons being composites of such hadrons as protons, neutrons and pions. Their nature varies more in comparison to the simple positive or negative charge of QED. For instance, such baryons as protons and neutrons are composites of three quarks of different color for the neutron to be internally neutral of parts. Apart from the nucleus, such mesons as muons are comprised of matter and antimatter comparable to positive and negative charge. As for maintaining the greater energy of the atomic nucleus at shorter range, it occurs by emitting gluons for the transformation of a baryon into a meson being comprised of a quark and anti-quark pair.

The Heisenberg uncertainty principle is a determining factor of infinities. The combined probability of determining an electron's whereabouts and momentum, for instance, equals the Plank Constant $\hbar = h/2\pi$. By interpretation, according to the perturbation theory, there is a probability of an electron interacting with a photon of nearly a zero wavelength for it to be equivalent to a particle of nearly infinite mass-energy. However, if such virtual energy is less detectable by the natural world in which we live, then renormalization is plausible.

Is quantum renormalization consistent with relativity theory?

By relativity theory, it takes an infinite amount of mass-energy for mass to obtain light speed. By quantum theory, a light wave nearly of zero length nearly equates to a particle of infinite mass-energy, which is undetectable according to the uncertainty of quantum probability. The two theories are consistent inasmuch as neither one of them allows the detection of a particle, either of light or matter, to be of infinite mass-energy. Their main difference is that of how they comply with different states of equilibrium, being of either gradual or quantum change.

11

QUANTUM WAVE MECHANICS

William Rowan Hamilton (1805–1865) formulated wave mechanics to combine wave properties with particle interaction. He noticed a similarity in form in mathematical formulations of the principles of least-time and least-action. Pierre de Fermat (1601–1665) offered the principle of least-time as a means to describe a light path. The least-action principle was formulated by Pierre-Louis Moreau de Maupertuis (1698–1759) for describing the dynamics of interaction between particles.

Properties of various physical media for the propagation of light waves are according to their refractive indexes, defined by the equation

$$\mu = \frac{c}{v}$$

The Greek letter μ denotes the refractive index of the medium, c is light speed in vacuum space (or ether), and v is its speed through a physical medium. The refractive index for vacuum space or ether is unity. It is greater than unity for other media. Since c is constant and since $\mu = c$, the speed of light is slower in a refractive index greater than one. The slower speed of c is justified by the conservation of momentum since the wave of increased inertia is of the denser medium. A speculative interpretation could be the denser medium converts light energy to more mass momentum of lesser speed momentum. However, except for the conditions of general relativity, it would have less kinetic energy except for observers in the field having less of it as well.

Fermat's principle of least time is according to the relation where time along the total path from point A to point B at each increment of distance ds through a medium of refractive index u is

$$\int_A^B u \cdot ds$$

A particular requirement of this integral is it needs to be an extreme, either as a maximum or a minimum value. The value of least time comes by way of associating u with c/v, such that the greater u results in a slower v for more time.

The shortest distance from point A to point C is a straight line, but the time can be faster from a diversion from point A to point B and then to point C if the distance from A to B is of a less dense medium.

The significance of this latter result with regard to the law of refraction is in realizing a total path that light takes through two media of two different indices. Light paths from A to B through the first medium and from B to C through the second medium need to be a minimum for the total time. The least time does not mean the shortest path; it rather depends on the angle of entry. The shortest path is a straight line from A to C, but the time along this path is greater for the medium having a greater refractive index. If the angle of entry is from a shorter path of the medium of a greater index, then the time can be less in the new and longer path. The law of refraction also expresses this same result. The least time principle is thus according to the law of refraction in view of wave theory.

The principle of least action is similarly expressed according to integration in the manner

$$\int_A^B mv \cdot ds$$

Momentum mv replaces refractive index μ of the medium. An increment of distance ds along some path from A to B thus has the momentum mv for action. For the total path from A to C through two media of two different indexes of reflections, the same condition exists for less action as it does for less time. The least action principle thus determines the path of a particle in a conservative field of force in the same manner the principle of least time determines the path of the light wave. A conservative field of force is one in which the total energy, potential and kinetic, stays the same.

The Hamiltonian action is a formulation of least action in view of a conservative field of force. A constant force in a homogeneous medium is simply the product of a particle's momentum mv and its distance moved r. Total energy before, during and after passing is the potential energy of a field in association with the particle in addition to its kinetic energy of relative motion. For constant field energy, Hamilton distinguished between the action of the particle and the total field energy as varying in time and place. The Hamiltonian action A thus takes the form A = S – Et. E denotes total energy, t is the time of action, and S is the Maupertuisian action mvr.

Anywhere along the path of the particle, or along the crossing of paths of a swarm of particles, the action varies with time. At any place where total positive and negative action is zero are the relations

$$Et = mvr$$

$$\frac{E}{mv} = \frac{r}{t} = v$$

An internal action of a field in equilibrium is thus describable.

Hamilton's wave mechanics are significant in describing the evolution of waves as multiple interactions of particles and manifestations of particles in an association with wave packets. A mathematical theory of wave packets was later developed by Lord Rayleigh (1842–1919). It foreshadowed the wave mechanics of de Broglie and Schrodinger that included Planck's constant h.

Quantum Wave Particles

The discovery of the quantum along with the success of wave theory for describing the behavior of light indicates light has a dual nature. It behaves partly as a particle and partly as continuous waves. Louis de Broglie (1892–1987) further theorized the nature of matter as a wave-particle duality as well with regard to his formulating theory according to special relativity, wave properties and the quantum. Accordingly, energy E and momentum P of a particle moving freely in space relate to the number σ of waves having frequency f and wavelength λ and equate with Planck's constant h in the manner

$$E = mc^2 = \frac{m_0 c^2}{\sqrt{1-\beta^2}} = hf$$

$$\vec{P} = m\vec{v} = \frac{m_0 \vec{v}}{\sqrt{1-\beta^2}} = \frac{hf}{\vec{V}} = \frac{h}{\vec{\lambda}} = h\vec{\sigma}$$

The arrows indicate the waves and particles move in the same direction.

V denotes the velocity of the waves. As with waves in general, it equates to the product of their wavelength λ and frequency f as

$$V = f\lambda$$

This equation, in relation to momentum and energy, further equates as

$$V = \lambda f = \left[\frac{E}{h}\right] \cdot \left[\frac{h}{P}\right] = \frac{E}{P} = \frac{mc^2}{mv} = \frac{c^2}{v}$$

The case of photons moving at speed c gives

$$V = \frac{c^2}{v=c} = \frac{c^2}{c} = c$$

Since special relativity stipulates no information of events can exceed light speed in vacuum, the de Broglie waves carry no momentum or energy for them not to transmit direct effect to the observable world.

De Broglie depicted an electron orbiting the nucleus of the atom as a superposing of waves creating effects similar to the beats of sound waves of the vibrating strings of a violin.

Wave equations relating musical harmony of musical instruments were developed in the eighteenth century. Particularly, Jean Le Rond d'Alembert (1717–1783) proposed, in 1746, a one-dimensional wave equation for a string-like vibration:

$$\frac{\partial^2 u}{\partial^2 x} = \frac{1}{v^2} \frac{\partial^2 u}{\partial^2 t^2}$$

The letter u represents the amplitude of waves as a function of energy, time and distance—energy being proportional to the square of the amplitude. The symbol ∂ represents partial differential, x its axis of direction in time t, and v its velocity. Although the equation is one dimensional, a plane is indicated by string-like vibration along a y-axis such that the average displacement is

$$\left(v\frac{\partial}{\partial x}+\frac{\partial}{\partial t}\right)\left(v\frac{\partial}{\partial x}-\frac{\partial}{\partial t}\right)y = 0 = \left(v^2\frac{\partial^2}{\partial x^2}-\frac{\partial^2}{\partial t^2}\right)y$$

The equation thus allows for solutions of internal action.

A three-dimensional wave equation was proposed three years later by Leonard Euler (1707–1783) with the symbol ∇ for three perpendicular axes. Erwin Schrodinger (1887–1961) also transformed de Broglie's wave-particle duality into a three-dimensional-space quantum wave mechanics in deriving more wave equations, such as a non-relativistic one for a particle moving in an electric field of the form

$$i\hbar\frac{\partial}{\partial t}\Psi(r,t) = \left[\frac{\hbar^2}{2u}\nabla^2 + V(r,t)\right]\psi(r,t)$$

ψ is the Schrodinger symbol for wave function, as for (r, t), ∇ is the Laplace symbol for three dimensional (x, y, z) coordinates, V represents total kinetic and potential energy according to Hamilton's conservation field rule, i is an imaginary number for the square root of -1, u is a reduced mass of a particle, and $\hbar$ is the Planck constant h divided by 2π. The equation is in compliance with conservation of energy according to classical theory.

Schrodinger's wave equation with quantum restrictions reinterprets the quantum conditions of the atom that was previously given by Bohr. Bohr had postulated a correspondence principle for including a wave condition where needed. Bohr's model further describes electrons circling the nucleus of the atom as being corpuscular in nature, whereas Schrodinger considered electrons as a diffused cloud whereby quantum restrictions result as certain nodal points of de Broglie standing waves.

A particular significance of Schrodinger's wave interpretation is Bohr's theory only explains refractive properties of atomic spectra as arising from an electron's change of orbit, as one of more energies closer to the nucleus, whereas refractive index of a medium for wave propagation explains why there is greater energy of attraction closer to the nucleus as a reflection of more energetic light. It provides explanation as to how the nucleus is able to contain more energy than the atom as a whole, even though the momentum of more energetic light as a particle could continue farther inward in manner of the atom itself being the medium of wave action.

Spin

Schrodinger's wave mechanics explained nearly all conditions of light spectra that Bohr's theory explained except for one that became explained with an assumption of spin as needed for explaining a particular spectrum. Bohr's theory explained it with the assumption of the magnetic moment of the electron being defined by its own spin, defined as half the value of the quantum constant $\hbar$. Nowadays it is simply said that the electron has one-half spin to that of photogenic energy, but the spin is symbolic in concept insofar as it does not compare with the spin of such a physical object as a ball.

George E. Uhlumbeck (1900–1988) along with Samuel A. Goudsmit (1902–1978) proposed a spinning electron to correct certain inadequacies of Bohr's theory to describe finer radiant spectra detected from the hydrogen atom. Although evidence supported the spinning electron explanation, there was no known reason for its existence except for Wolfgang Pauli (1900–1958) explaining it as consistent with wave equations presented in the form of a matrix allowing for an electron spin in either of two directions. When combined with the amplitude function ψ of waves, it further suggests polarization.

Charles Galton Darwin (1887–1962) then derived a wave equation that includes a polarization effect of the spinning electron in describing the fine structure of the hydrogen atom by means of a wave equation. Neither Pauli nor Darwin, however, derived wave equations consistent with the theory of relativity. This was a task taken on by Adrien Maurice Dirac (1902–1984) in deriving a relativistic form of Schrodinger's wave equations.

Schrodinger succeeded in deriving wave equations in classical form, as consistent with conservation of potential and kinetic energies:

$$\frac{1}{2m}\left(P_x^2 + P_y^2 + P_z^2\right) - K = 0$$

For wave interpretation, Schrodinger replaced momentum-vectors (P_x, P_y, P_z) with respective operators for wave interpretation of particle effects:

$$\frac{h}{2\pi i}\frac{\partial}{\partial x}, \frac{h}{2\pi i}\frac{\partial}{\partial y}, \frac{h}{2\pi i}\frac{\partial}{\partial z}$$

In addition, he replaced the energy K with the operator

$$\frac{-h}{2\pi i}\frac{\partial}{\partial t}$$

The operators above are only in compliance with conservation laws of classical mechanics in not including those of relativity theory. Schrodinger attempted a relativistic version, but he was unsatisfied with the result. Another version of theory was proposed by Walter Gordon (1893–1939) and Oskar Klein (1894–1977), but it also had reservations with regard to determining the probability density of a particular state of particle motion. The relativistic form of the wave equations requires symmetry for covariance, which was complicated because Schrodinger's equations in wave form indicated possible negative densities of matter (which was less acceptable by physicists at the time).

Negative values of relativistic invariance are still possible, but the form of the energy equation requires a more complex solution for including SRT conditions of combining energy with momentum according to invariance. In terms of three-dimensional space coordinates for momentum P, the invariant is

$$P_x^2 + P_y^2 + P_z^2 + m_o^2 c^2 = \frac{E^2}{c^2}$$

The significance of this invariance is that it is comparable to four dimensional spacetime in connecting to light speed and electromagnetism.

Including relativistic invariance in wave equations can unify theory, but the unification does not mean one theory becomes the same as the other; it means each becomes part of the other as parts of a more embracing theory. The new theory describes more phenomena, as exemplified by Dirac's inclusion of relativistic invariance in quantum wave mechanics.

Invariance allows calculation of unknown variables from known ones. Significant with regard to the derivation of relativistic invariance is the nullification of one factor canceling out another one. Derivation of Minkowski invariance does, in fact, consist of nullification by way of cancellation: The Lorentz transformations for coordinates of time t' and distance s' relate as ct' and s'; the transformation equation of s' is both subtracted and added to the ct' transformation equation. Results of added and subtracted values are then multiplied to become the invariant with some of the positive and negatives canceling each other. Wave theory also identifies with nullification in that waves of opposite phase cancel each other out in effect when they superimpose.

Wave interpretation provides insight into the physical natures of such nullification, which entails a different interpretation of the mathematics. A different commutative rule for multiplication, for instance, was pioneered by Hamilton and applied by Dirac for the application of his matrices in allowing Schrodinger's quantum wave mechanics to connect with special relativity theory. The commutative rule means A + B = B + A, as does AB = BA. To the contrary, A − B ≠ B − A is non-commutative, as 5 − 3 = 2 is positive and 3 − 5 = −2 is negative. Generally, 3 x 5 = 5 x 3 = 15 is commutative, but not according to Dirac's matrix rules.

Of particular significance to the non-commuting rule is its compliance with the number i as the square root of −1. In ordinary algebra the number one can either be $(1)^2$ or $(−1)^2$. The symmetry of Dirac's matrices, however, is such that $i^2 = −1$ has a unique interpretation: For every positive solution there is a negative or an imaginary solution as well. In reference to particle effects of electrons, for instance, the positive and negative factors represent opposite directions of counterclockwise or clockwise electron spin, and for antimatter to exist as an opposite state of matter.

Further significance is with regard to the Pythagorean theorem, which is a foundation of relativity theory. For instance, consider

$$(A + B)^2 = A^2 + AB + BA + B^2 = A^2 + 2AB + B^2$$

AB + BA = 2AB does not apply to Dirac matrices. Instead, BA is negative if BA is positive, such that A and B represent coefficients whereby AB + BA = 0 and $A^2 = B^2 = 1$ for light speed as unity. The non-commuting rule is thus a means of interpreting the positive and negative subtractions for deriving Minkowski invariance from Lorentz transformations.

Further significance is with regard to dimensional geometry. The matrices can increase to include four dimensional spacetime in compliance with relativity theory, but the particular significance of the non-commuting rule along with the number i for positive and negative values is what provides insight into a symmetry of relativistic covariance with regard to spin and antimatter.

This indication of antimatter is a square root of the momentum-energy equation being either positive or negative. It need not necessarily be either, as positive numbers each squared and added have only a positive value for the square root of their summation, but Dirac's matrices are of a symmetry consistent with the interpretation of Schrodinger's equations by Bohr, Born and Heisenberg whereby the wave aspect merely represents the probability of existence, and whereby Dirac's theory is verifiable as a test of predicting the positive and negative symmetry of nature.

Unity

An even greater significance of Dirac's approach is with regard to the unification of QED and relativity theory. This unification is in relation to spin. Bohr had accepted a suggestion to assume it applies as another condition of the electron in its orbit about the nucleus, as with a one-half spin of the electron to account for the empirical findings of a Stern-Gerlach experiment. It was discovered that a beam of atoms splits into a number of parts depending on the angular momentum of the atoms. This result indicated atoms split into three parts: -1, 0, 1, such that splitting the beam resulted in three parts, being positive, neutral and negative for half angular momentum in one direction balancing half angular momentum in the opposite direction.

Wolfgang Pauli (1900–1958) proposed in 1924 a new quantum degree of freedom with two possible values to explain observable effects of nuclear spectral. In 1926, he offered a 2 x 2 matrix for describing the two degrees of freedom in manner of non-relativistic spin of the electron. Shortly later, but independently, Dirac revealed a 4 x 4 matrix as a relativistic version of spin, as in accordance with the Minkowski invariant for energy and momentum. Dirac thus succeeded in uniting Quantum mechanics with special relativity, but not with general relativity in view of the singularity. However, there is possibility of including general relativity from its unification with special relativity.

Previously noted in this book is the gravitational escape speed being c' as one-half c instead of c. Its significance is that replacing c' for c in the Schwarzschild metric transforms it into a form of Lorentz invariance that further allows for the addition of velocities theorem to apply to gravity as well, even though it's a lot more mathematically complex.

There is further complexity with the unification of relativity with quantum wave theory, but there is historical perspective. Hamilton included a fraction one-half relating to three-dimensional rotation, as mathematically indicating it rotates consecutively in perpendicular directions, as subsequently one way and then the perpendicular way around the surface of a sphere for twice time and distance of spin. Similarly, a significant factor of a Fourier transform is the square root of one-half. Joseph Fourier (1768–1830) is noted for his mathematical analyses of heat transfer and his awareness of the greenhouse effect, but application of his mathematics applies to other areas of science as well. The square root of one-half, for instance, applies to amplitude A as a plane wave superposition in the manner

$$A(k) = \frac{1}{\sqrt{2}} e^{-(k-k_0)^2}$$

It relates to a wave-packet moving to the right and left according to a wave function w(k) = kv whereby u(x, t) equals both F(x ± vt) and further relates to localized disturbance of superimposed waves formed into a wave packet. Particular amounts of wave frequencies are needed to sustain a wave packet whereas destruction of any possible wave packet from it occurs instead. The result is consistent with Huygens's principle of every point of disturbance is the creation of more waves. Instead of waves being annihilated behind the front progression of wave action, they disperse from being a wave-packet.

The main reason here for pointing out the significance of the fraction one-half is for relating quantum wave mechanics to general relativity. The spin one-half of quantum mechanics relates as angular momentum. Similarly, gravitational potential, as in relation to orbital speed squared, is one-half the gravitational escape speed squared.

Light speed is a limiting condition of SRT by which matter cannot even reach. By GRT, the limiting condition is the escape speed instead of the orbital speed. For an addition of gravitational potentials theorem in analogy to the addition of velocities theorem, the limiting condition of the rotational speed is the square root of one-half c instead of c. The analogy is thus slightly more complex.

12

ÆTHER COSMOLOGY

Although Einstein stated that knowing how an absolute frame of reference can be detected is not necessary for the mathematical formulation of theory, he did acknowledge it could be useful for a better understanding of theory. Can an understanding of how an absolute reference frame is unobservant also be even more useful for further formulation of such theory as a unified one? Although it became generally accepted among established physics that mathematical formulation of evidence was only necessary, explaining it has still been considered useful by some physicists. One physicist of particular interest is Harold Aspden who began formulating his idea of a unified theory in the 1950s according to how an ethereal medium exists as part of its existence, whether of a plenum or partial vacuum.

Regarding conditions of the plenum noted by Descartes and others before, during and after his time, the only possible motion in it is circular, which constitutes extremely complex conditions of circular motion within circular motion. Consider, for instance, two spheres rotating in line where they touch each other. There is greater relative motion if both of them rotate clockwise than if they rotate clockwise and counter clockwise against each other, and various ways are necessary for continuation of motion and conservation of energy.

A variation of force even within an ethereal medium of a plenum is possible, but implications are that it is relative to how it is perceived, which further indicates the plenum itself exists in an extremely complex state of consciousness.

Can the perceived force be described in a manner relating to the laws of physics? Such geometrical conditions of mass volumes and force magnitudes seem essential. One physicist in particular, Carl R. Littmann, has shown how virtual particles of quantum physics combine in various ways relating magnitudes of size and mass in relation to circles and spheres.

Aspden had further formulated his theory with ideas of how such relations can interact in terms of gravity and electromagnetic forces. However, his theory was not accepted for publication in England supposedly because of a theorem published by Samuel Earnshaw (1805–1888) claiming the inverse square law is not possible of a solid crystallized state of ether. His theorem was published in 1842 in a journal of the Cambridge Philosophical Society. However, Aspden and previous theorists had offered a solution to the inverse square law as conditional to a crystal structure combined with a liquefied state of fluid motion that countered Earnshaw's theorem.

Mass and Volume Ratios

Arthur Stanley Eddington (1882–1944) speculated on the mathematical nature of the fine structure constant as an addition of numbers 1 through 16 that total as 136, being approximately one number less than the fine structure constant 137.015. For further speculation, multiplying 136 by three halves and then by nine results in the number 1836 that approximates to the proton mass in ratio to the electron mass, suggesting the creation of the proton from nine muons of 204 electron masses. However, the internal action of matter is not that simple. It is far more complex in consisting of internal actions between such virtual particles as various quark hyperons in the nuclei of atoms that perplexity influence the results in various ways. Even if nine muons combine to form a proton, they are considered virtual muons according to Aspden in the sense that energy lost in the process is not directly detected by observation. They thus have more electron mass before combining into other forms of energy.

Littmann spent decades analyzing mass and volume ratios of virtual energy particles. Much of his findings have been published in his **B**ooklet of Large and Small Mass Particles (For all Major Particle Groups in Physics).

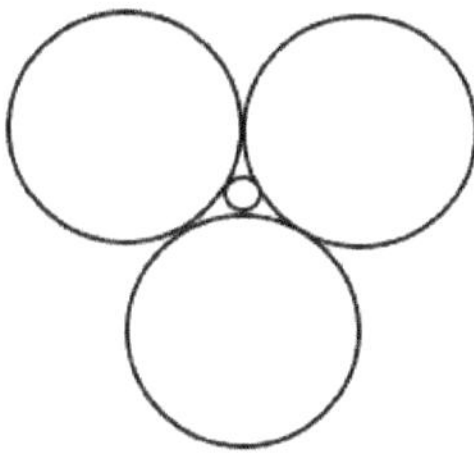

One of Littmann's simplest volume ratios with mass ratios is that of a pion mass in ratio to an electron mass. As depicted above, if three large spheres enclose a smaller one, the volume of a large sphere is 270.1 times larger than the smaller one. A pion mass on the average—with positive, negative or neutral charge—is empirically measured to be 270.13 electron masses.

If all the spheres above are contained within an even larger sphere, then the volume of the giant sphere is enough to contain 2702 electrons. A similar arrangement is of a giant sphere containing the alignment of four instead of only three larger spheres. Each of the four larger spheres are smaller than each of the three larger spheres, for each to be able to touch the tiny one in the middle, and the volume of the giant sphere enclosing the four larger ones is only large enough to contain 970 electrons. It is comparable to a particle referred to as the kaon. Noteworthy is that the two giant spheres together combine with enough volume to contain two protons: 2702 + 970 = 3672 = 1836 + 1836. If it truly exists as such regarding the hydrogen atom only being of one proton, then the proton exists as an average of the difference of combining forces.

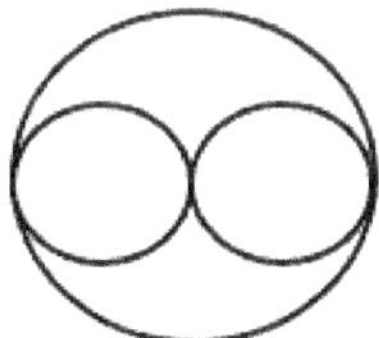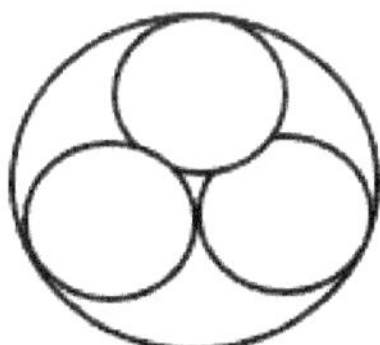

Another arrangement is of two large spheres each of a volume that is representative of a proton. One of them encloses two smaller spheres; the other encloses three smaller spheres. Each of the two smaller sphere volumes in ration to the proton mass is enough to contain 229.52 electrons. Each of the three smaller sphere volumes in ration to the proton mass is enough to contain 183.55 electrons. The average of 229.52 and 183.55 is 206.54. The empirical value of the muon mass, in comparison, is that of 206.77 electron masses.

This geometrical interpretation of mass in proportion to size is opposite to how physics theory interprets it. According to it, the proton of greater mass than the electron is the smaller of the two. Although it is about 1836 times more massive than the electron, it is contained in a smaller volume of a radius 1836 times less than that of the atom itself.

How, then, is it possible for 1836 electrons to be contained within a smaller volume for them to be about 1836 times denser to the fourth power? To speculate, the larger volumes are a means of containment. They consist of some source of virtual energy as wave packets in particular states of equilibrium. The electron is too weak to contain the proton, and it has an opposite charge of the proton for them to attract each other. It thus takes more energy to penetrate deep inside the atom to free the proton than it takes to free an electron.

Such speculation is consistent with such other conditions of physics as with regard to frequency. Shorter more frequent waves are needed to free an electron from the atom. The equilibrium condition of the atom is too slow to respond and maintain. Much shorter-frequent waves are needed to free the proton from the nucleus.

Further consider Boyle's law and the nature of the Planck constant. The latter maintains constancy according to a particular speed and a counter decrease or increase in the amount of mass being exactly the opposite of either an increase or decrease in mass radius. The former explains a change in density is to the third power as opposite to its surface pressure.

However, since the nucleus has about 1836 times more mass, its density is greater to the fourth power instead of just the third, which is consistent with the Stefan-Boltzmann law preceding the formulation of the Planck constant.

Constant frequency is also what distinguishes electromagnetic formula and the Plank constant from constant light speed. Regarding electromagnetic and Planck constants, they contain, in relation to the fine structure constant, velocity component 137.036 in ratio to light speed. Regarding the Planck constant in particular, it multiplied by light speed is 137.036 times greater than the electromagnetic constant. To speculate, energy waves need to move slower within the field of containment regarding their equilibrium states.

Further speculation of note is with regard to String theory and a greater nuclear density of the hydrogen atom being to the fourth power estimates closely to the fourth power of 2π, and 1836 divided by 137 estimates as $(2\pi)2$.

In relation to the internal mass energy of the electron, $m_e c^2$, or in relation to a photon's momentum, $m_e c$, Compton simply multiplied the Planck constant by a frequency or per wavelength in relation to a shorter orbital distance of the electron in the hydrogen atom. If the electron outside the nucleus of the atom is a cloud of a charge density towards the center of the atom, then it could neutralize with light penetration by being relatively denser at shorter radius by 137.036.

Similar to the Compton frequency in relation to internal mass-energy of the electron is that of Joseph John Thomson (1856–1919) having formulated it prior to Einstein's formulation of it. It became a fundamental part of Aspen's theory.

Aspen's Theory

A numerical progression theory has been suggested whereby the proton mass in ratio to the electron mass approximates to that of the fine structure constant multiplied first by three halves and then by nine. Geometrical ratios by Littmann indicate that the muon is part of the progression. However, nine muons contain slightly more mass than the proton-electron mass ratio. A possible remedy, theorized by Aspden, is that mass-energy is lost during the formulation of the muons into a proton. It is according to the fraction $2y^2/x^2$ whereby mass-energy is lost in ratio of length $(y + x)$ being reduced to length x.

Aspden's theory combines two formulas previously derived by Coulomb and Thomson:

$$\frac{-e^2}{(x+y)^2} = mv^2 \qquad \frac{2e^2}{3x} = mc^2$$

The Coulomb formula relates to a particular electrostatic interaction in relation to twice kinetic energy or potential energy whereas the Thomson formula refers to internal mass-energy. A particular difference is according to the extra length y of radius x.

The objective of the two formulas relates to how collision of two particles convert into internal mass-energy. In this case, $mv^2 = mc^2$, such that

$$\frac{e^2}{(x+y)^2} = \frac{2e^2}{3x^2}$$

$$\frac{e^2}{e^2} = \frac{2(x+y)^2}{3x^2}$$

$$\frac{3}{2} = \frac{(x+y)^2}{x^2}$$

$$\sqrt{\frac{3}{2}} = \frac{x+y}{x} = 1 + \frac{y}{x}$$

$$\sqrt{\frac{3}{2}} - 1 = \frac{y}{x} = 0.2247$$

$$\frac{y^2}{x^2} \cong 0.05051026$$

This numerical value subtracted from such an atomic mass as that of a muon relates to a loss of energy of muons forming into a more massive particle. The proton, for instance, is assumed to be formed by nine muons combining and losing energy to become the proton. Another creation is that of gravitons as a loss of energy from the interaction of two protons.

Proton Creation

Subtracting the numerical value of y^2/x^2 from 9 obtains 8.8989795. Multiplying that number by a muon of 206.3328 electrons instead of by 9 equates to the proton mass of 1836.1522 electrons.

To better understand how this result is possible, consider how particles of positive and negative charge combine. If five of them are of positive charge and if four of them are of negative charge, they can combine into three groups with two being of positive charge and the other of negative charge. The three groups can themselves combine into another particle of positive charge. By losing energy during the process is how the y^2/x^2 ratio applies to nine muons forming into a proton in the manner

$$\frac{m_p}{m_e} = \left\{9 - 2\left[\sqrt{\frac{3}{2}} - 1\right]^2\right\}\frac{m_u}{m_e}$$

$$\frac{m_p}{m_u} = (8.8989795)(206.3329) = 1836.1522$$

The ratio of the proton to the electron mass thus equals nine particles of reduced muon mass.

An empirical value of m_p/m_e as 1836.1527 is nearly the same as Aspden's 1836.1522, and his factor for the reduced muon mass is similar to another one of Littmann mass-volume relations in determining the ratio of proton mass to electron mass. As illustrated below on the left, a radii ratio 9.89898 to 1 is obtained of one of three large spheres closing around three small spheres. The volume ratio is 969.99912 to 1, which compares to an average kaon mass being either charged or not charged as a ratio 969.98/1.

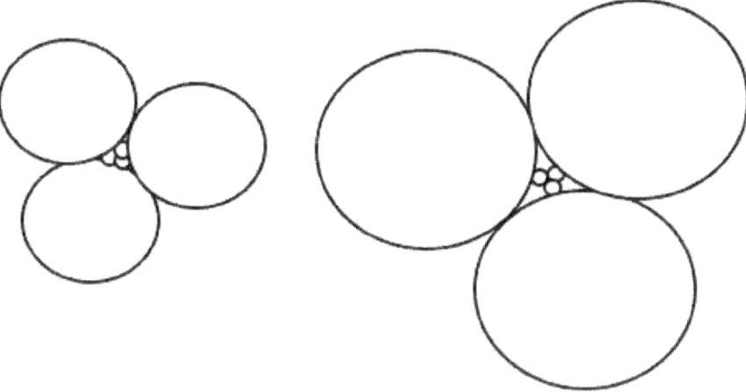

The illustration on the left is that of three small spheres packed between crevasses of three large spheres. The illustration on the right is that of three small spheres positioned opposite the crevasses. As for a ratio of large and small spheres on the left, it compares with the kaon having about 970 times more volume than the electron. As for volume ratios on the right, an outer sphere is about 2702 times greater than an inner one. The average of 970 + 2702 is 1836, as in approximate comparison to the empirical value of a proton mass of 1836 electron masses.

Greg Volk of the Natural Philosophy Alliance, which Littmann was a member, also geometrically determined the 9.89899 factor. Its determination is according to a tetrahedral pattern of four spheres apart from a center one. The center of the smaller one is the center of three perpendicular axes. The centers of the other four are at coordinate points (1, 1, 1), (1, –1, –1), (–1, 1, –1) and (–1, –1, 1). Distance D of the radius R from the origin is

$$D = \sqrt{1^2 + 1^2 + 1^2} = \sqrt{3}$$

The spheres are symmetrical to touch one another along planes of respective axes such that the radius R of a sphere is

$$R = \sqrt{1^2 + 1^2} = \sqrt{2}$$

Radius r of the inner square is coordinate distance D minus radius R, such that

$$\frac{r}{D} = \frac{D-R}{R} = \frac{\sqrt{3}-\sqrt{2}}{\sqrt{2}} = \sqrt{\frac{3}{2}} - 1$$

It squared is thus consistent with Aspden's derivation of the ratio of proton mass to electron mass.

In general, the proton is contained within the nucleus of the atom having a radius of about 1,836 times shorter than the atom. The theoretical existence of the ether does not vary in density, but it is reasonable that particles of higher density formed in it are contained in a smaller space whereby the larger one containing the electron is enough to contain the denser space-energy of the proton it surrounds. It requires electron-proton interaction be such that the electron energy combines with other energy to effectively counter that of the proton and have more separate effect for interacting with other forces.

Aspden Gravitons

Another use of the numerical factor of Aspen's theory is with regard to how gravitons and protons interrelate. Combining two protons results in what Aspden termed a tau graviton, which is similar in mass of a tau particle that is also included in Littmann's book. The empirical value of the tau is also given as 3477.19 electron masses.

Aspden's formula is of the form

$$\tau = \frac{2m_p}{m_e} = \left(1 - \left\{\sqrt{\frac{3}{2}} - 1\right\}^2\right) = 3485$$

It relates to a prior value before loss of energy from the interaction of two protons.

Regarding Aspden's gravitons, there is one of a particular charge acquiring lost energy from the interaction of two others from their opposite positive and negative charge values. During the process there exist 2τ and one g. Their interaction ratio of charge equates in the manner

$$\frac{\frac{2}{\tau^3}+\frac{1}{g^3}}{2\tau+g} = 3\,\frac{\frac{1}{g^3}}{g}$$

The right side of the equation pertains to volume charge of the undivided graviton in ratio to its total charge remaining the same. The equation simplifies in the manner

$$\frac{2g^4}{3\tau^3} + \frac{g}{3} = 2\tau + g$$

$$2g^4 + g\tau^3 = 6\tau^4 + 3g\tau^3$$

$$1 + \frac{\tau^3}{2g^3} = \frac{3\tau^4}{g^4} + \frac{3\tau^3}{2g^3}$$

$$1 = \frac{3\tau^4}{g^4} + \frac{\tau^3}{g^3}$$

Further simplification is by use of a computer. It was found that in terms of unit charge for all particles the mass ratios of τ and $\breve{g}$ to that of the electron are $\tau = (3487)m_e$ and $g = (5062.3)m_e$.

Aspden's formula further equates the gravitational constant G to electromagnetic charge e^2 in the manner

$$G = (16\pi^2)\left(\frac{1}{108\pi}\right)^6 \left(\frac{1}{g}\right)^8 \left(\frac{e^2}{m_e^2}\right)$$

$$= 6.67259 \times 10^{-8} \times \frac{cm^3}{(gm)(sec)^2}$$

The presently established numerical part of the G value is 6.7428(67). The numerical value obtained by Aspden's formula is significantly close considering various internal actions of tau particles combine to form gravitons being of a virtual nature.

Comparing Theory

How can a graviton have more mass than the proton that emits it?

Aspden's gravitons are not just emitted particles. They are part of the process of converting energy in maintaining the presence of such matter as the proton. The process in relation to the ether includes a G-frame and an E-frame in respective reference to gravity and electrostatics. They are supposedly reciprocal, but the process is complex.

Significant of the process is that it relates gravity to mass. It could further relate according to entropy whereby virtual energy converts to mass according to particular states of equilibrium.

As another part of the process, Aspden related it to a cubic structure of ether. How such a cubic structure of ether exists does not appear to be directly verifiable, but its effect is explainable according to geometrical difference in how spherical volumes can occupy space in contrast to how it theoretically can divide into cubic volumes.

Although the ether is somewhat speculative, Aspden's formulation of it is based on established theory. It relates to a relativistic modified version of Newton's inverse square law according to the condition of perihelion. In assuming gravitons propagate at light speed, gravitational action is delayed per distance. Einstein's general relativity equations entail the difference, which was verified in coinciding with the gravitation delay of planet Mercury around the Sun. Aspden further modified the effect according to a Gauss law formulated by Carl Friedrich Gauss (1777–1855). The law is complementary to Coulomb's inverse square one relating point charges of an electrostatic field of charge.

According to Gauss's formula, point charges of the electrostatic field are contained within surface areas of volumes for them to be in need of penetrating a particular thickness to escape. By applying part of this condition to Einstein's modified Newtonian formula, part of the new version was doubled in effect for further modification. A link was thus supposedly discovered for more accurately connecting gravity with electric charge.

The formulation is much more complex regarding how radii interrelate and so forth. Aspen's E-frame of electromagnetic charge does contain an ethereal particle he named the quon that has only repulsive effect. It equates to the electron in the manner

$$\frac{2m_0}{3r} = \frac{3m_e}{2d}$$

$$\frac{m_o}{m_e} = \frac{9r^2}{4d^2} = 0.04117589$$

The letter d denotes a side length of a cubic lattice of 27 quons. Rotation of the cubic lattice is more restrictive than that of a sphere. Its functional property includes the Gauss theorem whereby Aspden equates distance d to a charge density to the 27 lattice particles within the electron radius $d = (27)4\pi r_e$. The $4\pi r_e$ is that of the electron radius r_e according to the Thomson formula along with its 4π thickness according to the Gauss theorem. The number 27 is that of cubic lattice particles further relating to a force structure of the electron in terms of volume and geometrical symmetry for positive and negative charge.

Note: 9/4 compares to 3/2 in ratio to 2/3 being 1/3 + 1/3 charge of a quark in the nucleus of the atom in comparison to electron charge.

Radius r relates to the Planck constant h divided by $4\pi m_e c$. It is according to the momentum of an electron moving at twice light speed. The twice speed is that of systems approaching each other. Radius r thus relates to one-half the value of the Compton wavelength relating to a shorter perimeter of the hydrogen atom according to the fine structure constant for its quantum frequency to relate to $m_e c^2$.

All this and more variables for formulation is extremely complex, but the results are supportive of the previous results obtained by atomic and subatomic particles losing energy while combining their masses to become more massive ones. Eventually obtained are such relations as

$$\frac{r}{d} = 108\pi \left(\frac{8}{1843}\right)^{1/6}$$

$$\frac{hc}{2\pi e^2} = 108\pi \left(\frac{8}{1843}\right)^{1/6} = 137.0355915$$

$$G = (16\pi^2)\left(\frac{1}{108\pi}\right)^6 \left(\frac{1}{g}\right)^8 \left(\frac{e^2}{m_e^2}\right)$$

$$= 6.67259 \times 10^{-8} \times \frac{cm^3}{(gm)(sec)^2}$$

This gravitational formula result is thus consistent with previous theory regarding its historical development.

However, there are other ways to equate the gravitational constant according to a particular mass quantity for it to further equate to electromagnetic and quantum constants as well.

Gravity Electromagnetic and Quantum Theory

The ratio of two equilibrium states of light and mass are according to light speed multiplied by the fine structure constant: $c/137.036 = v$. It is also according to the electromagnetic unit squared per product of light speed and the reduced Planck constant: $e^2/\hbar c = v$. The equilibrium state of mass in ratio to that of light is thus a decreased unit of light speed.

As to how continual change of reality occurs, the equilibrium conditions further relate to relative motion and gravity in the manner

$$\frac{r}{d} = 108\pi \left(\frac{8}{1843}\right)^{1/6}$$

$$\frac{hc}{2\pi e^2} = 108\pi \left(\frac{8}{1843}\right)^{1/6} = 137.0355915$$

$$G = (16\pi^2)\left(\frac{1}{108\pi}\right)^6 \left(\frac{1}{g}\right)^8 \left(\frac{e^2}{m_e^2}\right)$$

$$= 6.67259 \times 10^{-8} \times \frac{cm^3}{(gm)(sec)^2}$$

They interrelate in relation to the fine structure constant.

Conditional to this mathematical formulation are quantum factors. Mass, velocity and radius parameters m, v^2 and r of e^2 represent a constant product similar to the parameters m, v and r of the reduced Planck constant $\hbar$. With v itself being a constant of no change, the product of parameters m and r is also constant, but varying in manner of either an increase or decrease of m being the reverse of that of r. Hence, the numerical product of grams and centimeters calculates in the manner

$$mr = \frac{\hbar}{v} \simeq \frac{1.05 \times 10^{-27}}{2.2 \times 10^8} \simeq 4 \times 10^{-36}$$

Calculation of m is according to the ratio

$$\frac{e^2/mr}{Gm/r} = \frac{e^2}{Gm^2} \cong 3.42 \times 10^{-12}$$

$$\frac{e^2}{G} \cong (3.42 \times 10^{-12})m^2$$

With m² being of unit value of grams squared, the actual value of m is about 1.85 x 10⁻⁶ grams.

The numerical value of r calculates in the manner

$$r = \frac{\hbar}{mv} \cong \frac{1.05 \times 10^{-27}}{(1.85 \times 10^{-6})(2.2 \times 10^{8})} cm \cong 2.58 \times 10^{-30} cm$$

Further calculation reveals

$$e^2 = mv^2 r \cong 23.07 \times 10^{-20} \, gm(cm)^3 \div (sec)^2$$

It is the same as $m_p v^2 r_n$.

The mathematical formulation began with the condition $Gm/r = e^2/mr$. A similar formulation with regard to the Planck constant $\hbar$ is

$$\frac{Gm^2}{r} = \frac{e^2}{r} = mv^2 = \frac{mc^2}{(137.036)^2}$$

$$\frac{Gm}{rc^2} \cong \frac{1}{(137.036)^2} = \frac{e^4}{\hbar^2 c^2}$$

There thus appears to be another particle of mass determining the value of the gravitational constant.

However, interpretation does not mean the graviton is more massive than the electron. A more feasible interpretation is that it relates to a greater amount of mass than just that of one electron. A similar condition, for instance, is how Compton related the internal mass-energy of the electron according to the Planck constant h. The electron does not move at light speed, but Compton related its momentum according to it in manner of the radius of the hydrogen atom being less according to the fine structure constant and multiplied by 2π. It thus relates to an orbital distance of a smaller atom. Dividing the distance, as wavelength, by light speed c, it relates as a frequency that, further multiplied by h, equals $m_e c^2$.

Gravitons are also assumed by theory to move at light speed. Their interrelation to the electromagnetic and quantum formulas is thus more complex and not in itself a complete unification of theory. However, an interpretation is provided at the end of the next chapter. It relates as a recycling process whereby mass is emitted as a virtual form of energy. The emitted energy is immediately replaced, but it also gradually converts back to a positive form of energy to fill all of space with the recycling process for vacuum effect. The recycling distance is the radius of the observable universe itself.

13

THE RELATIVITY OF HUBBLE COSMOLOGY

Why an infinite number of stars does not light up the sky at night is known as Olbers' paradox. Aristotle had proposed a finite number of stars. Kepler had considered the night sky indicates a finite universe. A century later, Edmun Halley (1656–1742) argued to the contrary, an uneven distribution of stars could allow for the appearance of a dark sky with stars too far away to be seen. In 1743, Jean Phillippe Loys de Cheseaus (1718–1751) considered the universe is finite unless space somehow absorbs starlight (as is also assumed by tired light theory). The debate continued. Heinrich Wilhelm Mauthaus Olbers (1768–1840) considered interstellar dust as an absorbing medium of space allowing for an infinite universe of infinite light.

The main criticism of the interstellar dust theory was the belief that dust in thermal equilibrium absorbs enough light from infinite stars to heat up and emit the light back out. However, Eddington and others analyzed there can still be a state of thermal equilibrium for some other process than that of just heating the medium if the overall temperature of the universe is only about three degrees Kelvin. According to tired light theory, for instance, space partially absorbs light energy. However, such theory is not presently established. Presently established instead is big bang theory assuming the universe is finite, as Einstein had assumed, and expanding. Its expansion is according to an interpretation of the Hubble constant by other physicists.

Tired light has been dismissed for several reasons. For one reason, how the collision of light with dust particles does not result in a distorted view of the cosmic background has been claimed to be in need of explanation. However, it can be argued, to the contrary, that an ethereal medium need not itself alter the visibility of the light source being the propagation of electromagnetic cycles, and there are many so-called dust particles throughout space along with opposing light moving in all various directions.

Another reason given for dismissal of tired light has been a Tolman brightness test supposedly distinguishing red shift effect in favor of recessional speed instead of by the absorption of light energy. However, the Tolman brightness test—in favor of big bang in its original form—is not totally supported by more recent astronomical data. Even though it remains a theoretical possibility, it has been modified to include dark energy allowing for an increased rate of expansion.

Big Bang

The birth nature of the universe in the form of crystallized matter from an expansion of energy was proposed by Robert Grosseteste (1175–1253) as early as 1225. The more modern version of cosmic expansion began in the 1910s when Vesto Slipher (1875–1969) began observing radial velocities of galaxies. Carl Wilhelm Wirtz (1876–1939) observed there is a red shift in light spectrum from more distant sources that is of less energy than a blue light spectrum. In 1929, Edwin P. Hubble (1889–1953) determined a systematic red shift per distance relationship. He and Milton Humason (1891–1972) then formulated the Hubble's law with regard to the red shift per distance of cosmic light sources. A plausible explanation of it is the Doppler principle of light spectrum affected by relative motion of its sources because of galaxies of stars receding from us as well as from each other. Another possible explanation is the tired light one of light-energy gradually being absorbed as it propagates through space.

Einstein had also contributed to the establishment of big bang theory by assuming the universe is finite and static instead of infinite in extent. A condition for it to be as such is that it needs a sufficient amount of mass to contain itself. Too much mass could cause it to contract; not enough mass could allow it to expand. He considered a repulsive force counter to gravity as a means to counter these two possibilities, which he theorized as a cosmological constant Λ inserted into his field equations for a modified form in the manner

$$R_{\mu\nu} - \frac{1}{2}Rg_{\mu\nu} - \Lambda_{\mu\nu} = \frac{8\pi G}{c^4}T_{\mu\nu}$$

The addition of constant Λ is a theoretical possibility allowed by the mathematical process of integration. Integration is a reverse process of differentiation whereby exponent powers become reduced in the manner x^3 becomes $2x^2$, x^2 becomes $1x^1$, and x becomes $0x^0 = 0$. Since differentiation eliminates constants, they become unknowns with regard to the reverse process of integration.

Alexander Friedmann (1888–1925) examined Einstein's field equations and asserted they indicate the finite-static-universe is unstable even with the cosmological constant. The discovery of a red shift in more distant starlight further indicates the universe could be expanding. Einstein concurred, referring to his insertion of the cosmological constant as his biggest blunder, but it has of late been reconsidered to explain dark energy as needed for an increased rate of expansion of the universe.

Although Einstein's insertion of the cosmological constant was not accepted by Friedmann, solutions to Einstein's field equations were derived with the intent of explaining an expanding universe. Georges Lemaitre (1894–1966) had suggested in 1927 that an expanding universe is traceable back to a time of origin. What developed is the now-standard FLRW metric that was named in honor of Friedmann and Lemaitre along with contributors Author Geoffrey Walker (1909–2001) and Howard Percy Robertson (1903–1961).

The FLRW metric is conceptually as well as mathematically complex. It includes a cosmological principle with regard to an observable expansion of the universe being conditional to homogeneity and isotropy. Conditional to it, for instance, is observers not actually located at the center of the universe still perceive themselves as relatively located at or near it, consistent with covariance of special and general relativity whereby observers in any system of inertial motion, either as relatively at rest or in a gravitational field, perceive the local measure of light speed the same as in gravitational free space.

Spacetime curvature of general relativity also applies whereby light paths curve for observers at the edge of the universe to perceive themselves to be near its center. As to how curvature maintains as the universe expands is more complex. Spacetime curvature of light within a larger universe of the same mass should be less according to theory, but there is another contributing factor regarding its expansion. Relativistic effects are inclusive of both gravity and relative motion. Expansion itself is thus a contributing factor in regard to relative motion.

How the universe now has enough mass to be a singularity is somewhat another paradox but not a contradiction. It is explainable similar to the clock paradox. Five billion years in the past it should have been much smaller and denser than it is now unless light energy has somehow been converting into matter, which is assumed to do according to a Higgs' mechanism proposed by Peter Higgs whereby a special Higgs' boson combines with other bosons of light to convert into mass particles.

Whether this singularity condition is a paradox or contradiction, theory does not appear to be complete. Perhaps the condition $c = 2c'$, derived according to the Schwarzschild metric, applies in a manner whereby the size of the universes is itself relative.

Expansion itself can be a relativistic factor. Light speed within a gravitational field is slowed in ratio to the field strength according to general relativity. As the universe expands and the field strength decreases, light speed farther from the center of the universe should increase according to theory. However, clocks and the measure of length become faster and longer as well. There can thus be relativistic nullification of effect.

The origin of the big bang is assumed a singularity expanding about 13.7 billion years ago from a dimensionless point space. Within an extremely minute second, the expansion rate inflated to a rate faster than light speed until it became comparable to a golf ball. Its rate of expansion then decreased, cooling to allow mass to form. Neutrons formed after a second to suddenly fill the universe. Further expansion of the universe has cooled it to its present average temperature of about 2.7 degrees Kelvin, as assumed, observed, and calculated in the 1950s.

Supposedly the universe was initially too hot for even light to shine. It did not shine until about 380,000 years to become a microwave background radiation of today with no common origin of location. The discovery of this radiation in the 1960s was what decided it as more preferable than steady state, which is another expansion theory whereby more stars and galaxies are created as other stars burn out of existence while the universe expands. It is evident they do burn out of existence while others are being created.

A timeline is an essential aspect of big bang theory. A simple explanation of it is twice distance equals twice the rate of recession. Assume a value of the Hubble constant H is 70 km/sec/mps. (The Hubble constant is a measure of light frequency per amount of distance at which the light source is from the observer.) Dividing 70 km/sec by a million parsecs (one parsec equals 30.9×10^{13} kilometers) equals 2.27×10^{-18} of something per time. If the something is one centimeter, then the result is centimeter per second. If c is an upper limit for the present Hubble distance R_u of the universe, then $R_u = c/H \approx 1.32 \times 10^{28}$ centimeters.

The expansion of the universe is assumed to have begun about 13.8 billion years ago, which is about 5,037 billion days, 120,888 billion hours, 7,253,280 billion minutes, or 435,196,800 billion seconds. Multiplying those billions of seconds by light speed of 3×10^{10} seconds per centimeters equals about 1.3×10^{28} centimeters, which is assumed to be the present radius of the visible universe. However, although R_u represents the radius of universe as the most distant light source from the observer relatively at its center, it is not the actual distance seen at the time of emission because of more time it takes light to reach the observer and for it to have enough energy for perception.

Tired Light

When Hubble and Humason formulated the Hubble constant in 1929, Fritz Zwicky (1898–1974) proposed an explanation in terms of tired light. It assumes light loses energy by collisions with mass particles along its journey through space. For instance, electrons of cosmic plasma could more easily detach from atoms than do electrons of ordinary matter. Light of electromagnetic propagation would then decrease in energy by inelastic collision. However, the Zwicky proposal was dismissed for two reasons: It had been believed to have failed because of evidence supporting the Tolman brightness test indicating big bang is consistent with the Doppler effect of relative motion, and because of the assertion that the spatial medium for tired light should cause a blurring condition regarding the observation of more distant stars. However, argument to the contrary of the second argument was that light propagating through an invisible medium of space as a constant change in effect in contrast to effects of visible particles throughout space can have various conditions to overcome whereby resulting effects are of an average variance according to the average density of ordinary matter in the universe.

A discovery of periodicity, or quantization, as it was interpreted, was announced in 1966 and 1967 by astronomer William G. Tifft (1932–2022). Other astronomers reported similar finds. What appears to occur is a greater Hubble constant value between more dense clusters of galaxies. An average density of mass in the universe could thus be a determining factor of the Hubble constant. More variance in effect is thus allowable. As for possible confirmation, a denser part of the universe absorbing more light energy should correlate with the missing mass of the universe as being more invisible.

The decrease in light energy is assumed, according to tired light, to be caused by cosmic plasma whereby a minute portion of light energy splits off to become the main source of the microwave cosmic background radiation that appears from every direction as having no common origin.

That all space contains a varying degree of plasma has historical support. Olaf Bernhard Birkland (1867–1917) explained as early as 1896 the origin of the auroras of northern and southern lights near the poles are due to upper plasma near the poles being more responsive to the colder weather conditions. Even though his theory was an extension of Maxell's electromagnetic theory, it was dismissed by Lord Kelvin claiming all space between Earth's upper atmosphere and the sun is void of matter. However, a magnetic field of plasma was discovered in 1967 by a military probe. Overwhelming evidence in support of the theory occurred by way of orbiting satellites in the 1970s.

Another possible explanation is consistent with wave theory insofar as shorter waves have more vibrating energy throughout space than do longer ones. The shorter wave vibration is even consistent with the tiny origin of the universe being equal to the present energy of the much larger universe inasmuch as greater action can occur between the greater density of shorter waves, and such greater mass-energy within smaller space is consistent with the Planck constant whereby the density increases to the fourth power, which is further consistent with the Stefan-Boltzmann law.

The Tolman Brightness Test

Although both big bang and tried light provide partial explanations for the Hubble constant, their explanations contrast with regard to further explanation of how relative motion of recession differs from continual decrease in light energy. Their main difference is with regard to the Tolman brightness test proposed by Richard C. Tolman (1881–1948) resulting from his and other astrophysicists' analyses.

Calculating the current position of a light source from where it is seen in the past involves a more complex determination of distance and size. For a light-source-speed-distance d of one-half ct at the time of emission, it takes starlight one-half more time to reach the observer. As seen, the actual distance d' of the source becomes three-halves more when seen: d + d' = ct/2 + ct/4 = (3/4)ct, and d'/d = (3/4)/(1/2) = 3/2.

The 3/2 ratio is also inclusive of tired light regarding a continual loss of light energy that decreases in ratio as the light energy itself decreases. At two different distances of two different light sources, with one being twice farther than the other, there is half as much decrease of light energy from the farther light source during the one-half shorter distance. Total decrease from the longer distance in comparison to the decrease from the shorter light source is thus 3/2 greater.

The 3/2 ratio is that of lesser light energy regarding both big bang and tired light. However, the results differ regarding calculation of exact location of light emission differing from the time at which they are seen. By recession, the source appears closer than it actually was at the time light was actually emitted because of the recession being extended during the time it takes light to reach the observer. Light intensity of the past appears more intense and the light source appears larger. Actual distance is thus calculable according to light intensity per size as well as its lesser frequency.

Dark Energy and Dark Matter

Observational evidence of this apparent difference in the perception of effect had seemed to be supportive of big bang theory. Its apparent prediction at the time was a reason for the acceptance of big bang instead of a static universe according to tired light theory. However, more recent determinations have indicated the universe now expands at a faster rate. An unexplained existence of dark energy has been assumed to exist as its cause. Along with dark energy, dark matter had been previously advocated in order to explain why some stars in spiral galaxies appear to orbit at a faster rate than the gravitational mass of the galaxy allows according to theory.

The presence of dark energy is now assumed to be the cause of the universe's increased rate of expansion. Dark matter has also been proposed to explain other effects. Overall, it is now assumed the universe has about sixty-eight percent dark energy, twenty-seven percent dark matter and five percent normal matter. The greater amount of dark energy is required for it to both overcome gravity and increase the rate of expansion because of a continual decrease in gravity with less density. Dark matter is needed to explain why spiral galaxies rotate faster than what their mass allows according to either Newtonian mechanics or relativity theory.

Whereas the dark energy proposal occurred from the astronomical observation in 1998 of distant supernovae indicating a slower rate of expansion in the past, dark matter is of a longer history, and it was assumed to explain why rotations of some galaxies are of speeds greater than their gravitational mass allows according to present theory. As early as 1939, Horace W. Babcock (1913–2003) noted a discrepancy in the rotational speed of the Andromeda galaxy. In the 1970s and 1980s, a general study of galaxies under the guidance of Vera Rubin (1929–2016) indicated rotational speeds of spiral galaxies generally do not appear to decrease according to Newton's inverse square law. Mordehai Milgrom proposed a modification of the law in 1983, referred to as MOND. In 2004. Jacob Bekenstein (1947–2015) offered a relativistic version of the modification according to space-time curvature. In 2008, John W. Moffat *Reinventing Gravioty* regarding modification of gravity (MOG) whereby the gravitational constant relatively increases farther away from the center of mass due to greater decrease in repulsive force away from the center of mass.

The existence of dark energy is not yet acceptedly explained. Einstein's cosmological constant has been reconsidered as a counter force to gravity that can somehow increase in effect as the universe expands, which would be consistent with modified version of gravity allowing less resistance to relative motion farther away from the center of the field. However, the amount of repulsive force needed has not been determined in accordance with empirical data and established theory to be enough for increasing the expansion rate. A dark energy of about 10^{122} times greater than what the cosmological constant provides has been claimed by some physicists to be needed.

Modification of theory has been considered for creation of dark energy. A possible one to consider is with regard to the relativity conditions of black holes. With light speed being slower within a gravitational field, a loss of energy from its slower speed could be stored in another form. This other form of energy could be the repulsive force for explanation. As to how the energy is not seen other than by its increased repulsive effect, it could be because of the relativity of the black hole not only complies with covariance; its size increases along with an increase in mass. If mass is still being created by means of the Higgs' mechanism, for instance, then the universe could be expanding because of it as well, which is also a possible means of explaining the paradox of how the rotational speed at the edge of the universe does not decrease because of its expansion.

How this covariance is possible is with regard to how the black hole itself can differ. Suppose cosmic covariance is an effect whereby we perceive ourselves to be at the center of an observable universe. Suppose we then enter a black hole within the same universe. Since clocks become slower within the smaller and denser black hole, the universe outside it becomes relatively smaller in contrast to light speed being faster, even appearing not to exist while the black hole within it becomes a relatively larger universe to those of us in it. However, the smaller black hole universe expands according to the cosmological constant likewise increasing with the absorption of more incoming energy becoming mass and dark energy.

How can size determine the degree of relativistic effect?

Smaller black holes can be relatively denser in mass-energy. By absorbing a relatively denser amount of light energy that they contract, there is more of it per volume of space for a cosmological constant of a greater magnitude. If the cosmological constant also determines the relativistic contraction of size, then size itself could also be relative.

Explanation of dark matter is more speculative regarding conditions of measurement according to theory of reference. According to big bang theory the longer distance of the light source, according to the redshift spectrum, was a shorter distance of the past. The intensity of the light source decreases to the fourth power per distance and also according to the rate of accession. The latter is according to frequency and redshift. To the contrary is tired light whereby past distance is the same as the present one. Its size is thus relatively larger, in observational comparison, for it to either have more mass or less density of mass. The increase in mass per distance is an eight times ratio in contrast to the same gravitational potential being according to the ratio of the same increase in mass per same increase in its radius.

An accurate determination of the exact value of the Hubble constant has been difficult. Different observations and their determinations have varied over the years. Telescopes have only been able to detect observable starlight at limited distances. More recently, however, telescopes in spaceships have been able to view much more starlight than those on Earth's surface and even of the Hubble telescope in orbit. A particular telescope transported out into space is the James Webb Space Telescope, JWST. According to astrophysicist Stuart Clark, it was built to detect infrared light of longer wavelength in manner for us to see more with our eyes. Such detection has been the apparent presence of dust particles in intervening galaxies, supermassive black holes in centers of galaxies, larger galaxies and stars father away, and so forth. Study and assessment of data is still ongoing.

Values of Physical Parameter

Gravitational Constant: $G = 6.67428(67) \times 10^{-8}$ cm^3/(gm)(sec)^2

Proton mass: $m_p = 1.67262137(13) \times 10^{-24}$ gm

Electron mass: $m_e = 9.10938215(45) \times 10^{-28}$ gm

Light speed c: $2.99792458 \times 10^{10}$ cm/sec

Bohr radius of hydrogen atom: $r_a = 5.2917721067 \times 10^{-9}$ cm

Electron radius: $r_e = 2.8179402894 \times 10^{-13}$ cm

Neutron radius: $r_n = 2.881991 \times 10^{-12}$ cm

Fine Structure Constant: $\delta = 1/137.036 = e^2/\hbar c = v/c$

Planck constant: $h = 6.62667158$ (gm)(cm)^2/sec

Electromagnetic constant squared: $e^2 = 23.07 \times 10^{-20}$ (gm)(cm)/(sec)^2

Relating the Hubble Constant

Friedmann provided a gravitational formula relating the red shift to the mass density of the universe. The red shift is also considered here as a radiation absorbed by the medium of space for maintaining gravitational effect as vacuum effect.

For the vacuum effect to be possible, a positive outside force is needed of an equilibrium state whereby virtual energy converts into mass of atomic particles that in turn convert mass-momentum into the virtual particles of gravitons in compliance with entropy. There is thus a recycling process. The emitted radiation is a long-range and partly elusive process determined according to the probability condition of quantum physics typical of Earth's detection of only a minute portion of neutrinos passing through Earth.

Friedmann's formula is according to the average mass density of the universe in the manner

$$\rho \cong \frac{3H_1^2}{8\pi G} = \frac{3c^2}{8\pi R_u^2 G} = \frac{3(2GM_u)}{8\pi R_u^3} = \frac{M}{\frac{4}{3}\pi R_u^3}$$

In general, H simply equates to light speed c per radius such that HR = c with R being the radius of the universe. In relation to density $3R^3/4\pi$, c^2 further equates as the gravitational escape speed of the universe. Hence, c/R = H and $c^2/R^2 = H^2$, such that

$$\frac{2GM}{R} = c^2 \cong H^2R^2$$

$$\rho \cong \frac{3H^2}{8\pi G}$$

The value of H thus depends on the relative sized of the universe and vice versa. Its general form is also a means of comparing other formula derived according various other conditions of observation.

As part of the density formula is the simple equation HR = c. Another simple formula approximating H nearly to its more recent value from observation is according to the gravitational constant G divided by light speed c in manner of it relating to centripetal acceleration according to unit mass m and unit radius r squared per light speed c:

$$H_2 \cong \frac{Gm}{r^2c} = \frac{v^2}{rc} = 2.228 \times 10^{-18} sec^{-1}$$

Although the numerical value of G depends on the chosen units of measure, distance lengths cancel out in terms of centripetal acceleration per light speed for the result to relate as frequency: v/c eliminates one radius and one time unit of v^2 and r eliminates the other radius of the other v. It thus becomes per frequency.

As for the mass unit of measure determining the numerical value of the gravitational constant, it also cancels out regarding either centripetal acceleration or centripetal force. According to the former, $Gm/r = v^2$, the value of v remains the same regardless of the unit value of m. According to the latter, $Gm^2/r = mv^2$, the cancellation occurs because of the momentum of light being measured the same as the momentum of mass, in view of a change in light energy, providing the Hubble constant and gravitational mass are measured according to the same mass unit.

The Hubble constant also appears to interrelate electromagnetic and quantum theories in the manner of it per light speed at a distance equal to the diameter of the nuclear radius simply equating to the ratio of gravitational and electromagnetic forces between the proton and electron in the manner

$$\frac{H_3 r_n}{c\prime} = \frac{H_3(2r_n)}{c} \cong \frac{Gm_p m_e}{e^2} = 4.4 \times 10^{-40}$$

Converting H_3/c to $1/R_u$ and dividing both sides of the equation by r_n, instead of $2r_n$, renders a value for R_u as 1.35×10^{28} centimeters, which is the same as determined for expansion about 13.8 billion years ago.

Another relation, indicated at the end of the last chapter, is that of a unit mass m equating the gravitational formula to the electromagnetic one. It further relates in the manner

$$\frac{H_3 R_u}{c} = \frac{Gm^2}{e^2} = 1$$

A virtual form of m thus recycles back into a repulsive form after a distance equal to the radius of the observable universe.

This equating to unity could be significant regarding interrelation between gravity, electromagnetism and the Hubble constant. Questionable is the slight difference in values H2 and H3. Although their difference is slight, with the former being only about 7 x 10-20 difference in value, they both need to change the same in order to maintain equivalence with other factors. Some other factor of change need thus be applicable. A reasonable possibility is that of temperature. Boyle's law is according to a particular temperature. Avogadro explained the ratios of whole number combos according to the same number of molecules of particular temperature and weight. Derivation of the Planck constant included difference in temperature. The measure of such constants as G and H are inclusive of temperature as well. If values of both G and e2 have both been determined according to about the same temperature, then it is possible H2 and H3 do equate if both values of G and e2 are redetermined according to an absolute zero temperature. An example is that of a different observation of distant stars is determined from a spaceship farther away from Earth and the Sun that is also at a different temperature.

As for explanation of the black hole condition opposing universe expansion paradox, more complete theory could be required. Our observable part of the universe is assumed to be finite. It could be a black hole expanding with absorption of more mass. Another possible factor is that gravity can cancel itself out. In being consistent with entropy, if mass-density is the same throughout infinite space, then its existence is nullified. For conservation of entropy, variance in density need be constant, but there could be exchange in ways that mass outside the observable part decreases the black hole condition in allowing energy to escape in manner of Hawking radiation.

14

MASS MOTION ATTRACTION AND EQUILIBRIUM STATES

In the previous chapter, gravity was associated with the Hubble constant insofar as a minute decrease in radiant energy allows for a long-range effect of a relatively weak force of gravity per mass in comparison to such other forces of nature as atomic and electromagnetic. Further explanation is with regard to an integral part of quantum physics being a vacuum state of virtual particles as a field of virtual energy. Such virtual particles as gravitons are explainable as vacuum effect from the wake of mass-energy being both emitted and restored as a positive form of momenta of virtual energy evenly distributed throughout an otherwise vacuum space effect. The presence of mass interacts with itself to emit and absorb particular forms of radiation while maintaining in various states of equilibrium.

A particular part of the explanation is in relation to frequency. Gravitational waves are less energetic than those of electric charge because they more gradually change per distance to the positive form of repulsive energy. There is thus longer range of recycling effect through the medium of space for the repulsive form to already be distributed throughout spacetime for immediate effect.

The explanation is also consistent with tired light theory that has been accused of not having a viable explanation as to how light can decrease in energy and still allow the visibility of distant stars. How this visibility is possible is further explained as well along with its difference in effect of electrostatic charge. However, light energy being absorbed in space can also result in the expansion of the universe as well.

Further explanation includes a more in-depth analysis of the method of radiation superimposing to form observable mass in a manner consistent with how it relates to both relative motion and gravity. Mass is to be considered as the superimposition of wave packets of radiation whereby they comply with entropy and covariance of both relative motion and gravity to maintain equilibrium states of mass and energy by means of virtual energy.

Virtual energy is indirect effect, being directly invisible. Casual explanation as to how virtual particles cause attraction was considered unnecessary according to some physicists, such as Feynman. Mathematical consistency was only required by him, whereas explanation in in this book is more causal in support of such concepts as zero-point energy (ZPE) that Plank later proposed as a modification of his original formulation of the quantum.

All in all, visibility of stars is explainable along with gravity and electromagnetic attraction and repulsion. Also inclusive is such other explanation as that of a Casimir effect of attraction related to the closeness of metal walls, and of a geometrical description of mass as wave packets proposed for possible explanation as to how gravity, electromagnetism and the quantum nature of reality interrelate.

Virtual Vacuum Space

Whether space is only partially filled or is a plenum, quantum theory now describes vacuum space as containing virtual energy particles according to the Heisenberg uncertainty principle. Such virtual particles as gluons are assumed in order to explain observable effects that do not otherwise comply with predictions of quantum theory. A gluon is confined as part of a proton or neutron whereby it cannot be observed directly as an individual particle apart from a proton or neutron. It is verifiable only as indirect effect according to mathematical probability. It thus seemingly exists as a virtual particle. In general, the vacuum of space is now assumed to contain numerous assortments of virtual particles.

This quantum vacuum condition is not here contested; it is to include nonquantum conditions of continuous change in motion as well as discrete quantum effects. Matter at rest absorbs and emits discrete units of electromagnetic radiation as quanta, but quanta can also vary according to relative motion, gravity and electric charge in compliance with the relativity of covariance. For a more complete description of reality, relative motion, gravity and electric charge all comply with the Doppler principle as to how continuous nature of change in effect as well as a discrete quantum effect of change occur. The quantum is explainable as different equilibrium conditions consistent with both the constancy of light speed and the constancy of the fine structure speed that is about 137.036 times less than light speed. The latter being less than light speed simply complies with entropy whereby equilibrium conditions exist for a particular mass structure that slows the process of conversion through it in manner of still being conditional to covariance effects of relative motion, gravity and electrostatics.

As to how these equilibrium states can differ, light speed can be of a primary equilibrium state whereas the fine structure constant is a secondary one formed from the interaction between mass particles that are formed as part of the primary one. By entropy, change occurs according to disorder. The interaction between different equilibrium states is thus how the reality of change can occur.

ZPE

After Planck derived his quantum formula for resolving the infinity problem of black body radiation, he attempted to explain continuous change in force according to a concept of zero-point energy. He proposed to modify quantum theory for it to comply with the continuous changes occurring of relative motion and so forth. His effort was continued with proposed causal explanations of the wave-particle paradox by De-Broglie and a hidden variable approach by David Bohm (1917–1992). A stochastic interpretation of quantum probability conditions was also proposed by Jean-Pierre Vigier (1920–2004).

Although Planck had revolutionized physics in the year 1900 with the introduction of the quantum as a solution to an infinity paradox of black-body radiation, he did not accept some of its implications. He continued to pursue a more consistent solution with classical electrostatics. He contrived a possible solution in 1911 assuming quantum effects are particular oscillation modes of atoms. However, his assumption contrasted with Bohr's atomic theory whereby quantum jumps of discreet energy occur with absorption of radiation as well as its emission. In effect, the continuous manner of change in the relative motion of mass only occurs by reflecting radiation instead of by either absorbing or emitting it. However, a particular aspect of Planck's revised theory did receive recognition.

Plank added a term of one-half quantum frequency to his original equation relating energy of radiation to absolute temperature. He referred to this one-half quantum frequency as zero-point energy of an oscillator. Walther Nerst (1864–1941), who had formulated the third law of thermodynamics, interpreted the term in consideration of a possible heat death due to a loss of radiation emitted out of the universe, which could apply to an observable finite part of an infinite universe. He compared the one-half quantum frequency to the Boltzmann constant temperature T. Further consideration of Plank's additional term became evident in view of Heisenberg's uncertainty principle in that otherwise zero energy at an absolute zero temperature contradicts the condition of no possible determination of exact energy for any particular time.

Although the quantum is a discreet unit of change, it complies with the relativity of covariance. Such covariance also complies with gravity according to the slowing of light speed in a gravitational field. Quantum units were also determined according to the Avogadro constant being conditional to a particular temperature and a particular weight. Continuous change can thus be merely considered as part of the manner of maintaining the equilibrium state of the quantum.

Actually, the quantum is merely a particular equilibrium state of matter that is allowed to vary according to relative motion, temperature and so forth because of varying energy of light photons varying in their interaction with the quantum state of matter. The issue is thus of merely explaining the quantum state as another equilibrium state of existence.

The Casimir Effect

More support for ZPE is an explanation of the Casimir effect. It was proposed in 1947 by Hendrik B. G. Casimir (1909–2000). He and Dirk Potter (1919–2001) experimented with metallic plates for the measure of the Van der Waals force between the plates and their polarized molecules. They discovered an attractive force between the plates if they were close enough to each other.

Niels Bohr suggested to Casimir the experiments could relate to ZPE. Casimir formulated in 1948 a theory in compliance with Bohr's suggestion. Experimental evidence as accurate to within fifteen percent of the predicted value of theory was claimed in 1997 by Steve K. Lamoreaux. More accurate claims followed. However, exact results require experimental conditions to the extreme, such as exact vacuum conditions and perfectly smooth walls.

The Casimir effect relates to ZPE in manner of the Heisenberg uncertainty principle as variable conditions occurring of the quantum. The effect, for instance, is explainable as a polarization of positive and negative charge. Since like poles repel and opposite poles attract, bar magnets tend to align with opposite charged poles facing each other. The closeness of the walls is simply such a response further explaining how the quantum state can vary in effect.

Wave Packets

A wave packet consisting of the composition of various wave frequencies was proposed by Schrodinger in his formulation of wave mechanics. It was later interpreted as a probability condition of possible interactions of virtual particles. A wave packet is basically a localized disturbance that results from the sum of many different actions.

Superimposing of wave action as the conversion of energy could result as infinite energy within a nearly infinitesimal space, being consistent with big bang theory, except for the uncertainty principle requiring an infinite magnitude of energy has zero probability of detection. This uncertainty might be actual or merely invisible regarding its detection.

Such invisibility of entropy is typical of wave action. Waves superimpose to produce visible effect only if the medium of wave action changes in a way it can be seen. If action within a medium is counterbalanced, then a direct change occurring within the interaction need not be seen beyond it.

A connection between quantum and continuous change is with regard to the wave-particle paradox. Although Einstein explained the photoelectric effect of particle ejection from metals as particle effects of photons, further experimental evidence of interference is supportive of a wave interpretation of light because of the probability of one particle passing through one of two holes resulting in an effect on a screen consistent with superimposing effect of waves.

Einstein also referred to the photon as a wave packet. He even attempted to resolve a so-called wave-particle paradox in speculating that wave packets contain a minute particle within them, and wave effects do occur from particles of matter in particular states of equilibrium.

Light waves of higher frequency can free more electrons than does the same energy of more intense light waves of less frequency if the ones of less frequency convert part of their energy as molecular motion of heat. Explanation is analogous to microwaves of a microwave ovens being of much lower energy radiation than, for instance, ultraviolet waves. The cause is simply that waves of higher frequency allow less time for the equilibrium state of the particle effects to respond.

Einstein later offered an explanation of photons guided by waves. Waves can superimpose to be invisible to each other, but that does not conclude they do not lose energy while moving through space in manner of maintaining the visibility of their light sources. It is possible light energy can propagate evenly in perpendicular directions without interfering with each other for no change in direction to occur of them except by gravity and other factors of observable effect.

De Broglie was another wave-particle proponent in considering particle effects result similar to how sound occurs from overlapping of waves. Schrodinger developed a similar idea in his attempt to unite his wave mechanics with relativity theory. Moreover, in 1954, Bohm and Vigier mathematically developed a casual wave-particle duality explanation that became disregarded in favor of the Heisenberg uncertainty principle.

The primary distinction between electromagnetic waves as either light or particles of mass is that of the former can superimpose to occupy the same space whereas the latter cannot. To the extreme, light can be invisible to other light. However, by the Higgs mechanism, light contains momentum and converts to mass in increasing the speed of mass by mass either absorbing it by inelastic collision or partly absorbing and partly reflecting it. Various inelastic effects occur somewhere between these two extremes. Rather than light being reflected straight back from a metal, it can be partially converted to heat, or it can result in an emission of electrons if the light frequency is too fast for electrons to absorb it. To the extreme is the principal condition whereby mass emits fully what it absorbs. By theory, it absorbs and emits it as quanta, but what a system absorbs and emits also depends on particular temperatures as determined by Kirchhoff and other physicists in the nineteenth century. Relative motion is another factor according to the Doppler effect.

Contrary to quantum restrictions is the relativity of partially absorbing some of the light energy reflected in order for mass to not equal or exceed light speed. It is thus reasonable that light energy can convert to mass energy as well as being a particular source of quantum emission.

Such effects are part of standard model of how such elementary particles as electrons are created. There is also such antimatter as positrons opposite to electrons whereby the opposites combine to transform into non-massive energy that could possibly be either light bosons or virtual energy. For conservation of mass-energy, it is necessary for light or virtual energy to somehow recycle back to material particles. A particular theory with many contributors is the Higgs mechanism. According to it, the Higgs field provides mass to massless particles interacting within it. The result has led to unification of the electromagnetic weak force with both the strong force and relative motion, but not with gravity.

Distant Visibility

In relation to the Feynman diagrams, there are particles responsible for attraction and repulsion. The influence of virtual particles results in both the repulsion of electrons from electrons and protons from protons and the attraction between electrons to protons. Various effects also arise. For instance, two parallel wires, as Ampere discovered, contract if they both have electric currents flowing in the same direction and repel if the currents flow in opposite directions. Such effect is explainable in manner consistent with both Feynman diagrams of interaction between atomic or virtual particles and conservation of momentum. Particles are here considered as discrete units of wave action being consistent with both quantum effects and those of relative motion.

Accordingly, an electric current through a wire emits virtual particles having more momenta in the direction of flow than perpendicular to it and in the direction of two wires relatively at rest. The emitted virtual particles, in turn, collide to emit secondary virtual particles with less energy because they move partly in the same direction. There is thus relatively less momentum of energy difference moving in perpendicular directions between wires. There is a vacuum effect of attraction occurring. In contrast, virtual particles from currents moving in opposite directions collide more energetically for virtual particles to have more momentum difference in perpendicular directions between wires for a repulsive field of action created instead. However, it is also possible the frequency of wave action, regarding difference in superimposing effect, can result in either attraction or repulsion as well.

Regard this explanation as fundamental on a primary level from which attraction and repulsion become functional on subatomic and atomic levels. What results is a method of bonding whereby magnets can either combine to become one large magnet or disperse into tinier magnets in number.

The reason here for explaining electromagnetism is with regard to the need of explaining the visibility of distant stars according to a tired light theory as part of big bang theory. An analogy for the explanation is with regard to television. How is it possible that images are transmitted through wires for countless viewers to clearly see? The explanation is with regard to bar magnets providing direction of energy by dividing and duplicating into multiple magnets. Although smaller magnets are of less energy, their quantum density of energy maintains and expands for observable effect with the provision of additional energy source.

The video of television consists of the collection of images by cameras transmitted as electromagnetic signals propagating through both space and wires (or, more clearly, through cables). Required is a transmission of signals to be consistent with how the human brain distinguishes the data it receives. In general, a visible image maintains in the human brain for only a tenth of a second for each pictorial change given by the light source. Ten different images per second are what the brain comprehends with regard to a continuous sequence of scenery. In practice, between twenty-five and thirty complete pictures per second are received by the brain, with each picture divided into about two hundred thousand elements, or pixels. About two million individual details are thus perceived in total by the brain per second.

For transmission of data to be consistent with perceptual ability of the brain, the signals need to be within a frequency range as bandwidth. Entailed in the process are antennas and transformers to both transmit and receive. The signals are somehow amplified and multiplied for numerous observers. Signals themselves are electromagnetic waves as parts of an electric current surrounded by an induced magnetic field. How all individual images can be maintained is explainable if we consider the magnetic field is divisible into a field of numerous parts similar to how a bar magnet divides into smaller bar magnets. Packets of bar magnets of particular images are thus amplified in number in being duplicated for alternate routes. Moreover, waves can be of gradual change as well as quantum change.

Although the analogy only applies in the sense that waves gradually divide in all directions forward in similar manner of how bar magnets divide, there could still be a connection. A gradual change in the wavelength of light is possible by it having both continuous differences in reflection and a quantum difference in atomic absorption. The various differences are with regard to such different effects as that of relative motion and temperature.

Significantly, individual arrangements of bar magnets do not change as long as their medium of propagation is in balance, even though their energy for propagation either reduces or is amplified. Maxwell explained the process of electromagnetic propagation as conditional to space alone, separate from a physical medium. Starlight being electromagnetic waves of energy can thus exist as bar magnet packets that maintain individual images while losing energy as they interact with the virtual field of energy by which they advance as wave action through space. What remains to be determined is whether starlight itself is amplified for revealing more or less detailed information of individual stars in similar manner of the broadcast of television.

Variable Light Mass Gravity and Relative Motion

Einstein explained gravity as following spacetime curvature due to the presence of mass. What is mass, light and their difference? The main difference between them is that light superimposes with other light to occupy the same exact space. However, the measure of space is relative to whatever measures it. Einstein's mass-energy equation indicates mass is energy per light speed squared. It would become infinite mass-energy at light speed except for the needed amount of infinite energy not being attainable as such. Mass also varies in speed, but the speed of light in the virtual vacuum of gravitational free space is assumed constant. However, light speed relatively decreases in a massive medium and in a gravitational field. It is even assumed to convert into matter by means of the Higgs mechanism. Further assumed here is that light both superimposes and reflects from matter, and that matter converts back to light as a reverse process of the Higgs mechanism.

A Pauli exclusion principle refers to a condition whereby mass particles named fermions cannot occupy the same space that other fermions occupy. To the contrary is electromagnetic radiation of such boson particles as photons superimposing on other bosons according to a probability condition of quantum wave theory. As to how such opposite conditions coexist, a so-called quon algebra has been proposed as a possible method of allowing for slight changes to occur between extremes of these two principles in view of a wave-particle duality possessing both light and matter properties. For instance, J. W. Goodison and D. J. Toms proposed in 1993 that particle creation and annihilation in the quantum vacuum state of virtual particles is allowed by different energy densities for interpolation according to the quon algebra.

The quon algebra is similar to previous methods of formulation prior to mainstream physicists regarding ether as unscientific, even if it does exist, because of its invisibility not allowing for empirical confirmation. Einstein excluded it as unnecessary for his formulation of special relativity theory, but he suggested it could be useful for a better understanding of theory. Even though casual explanation is controversial, it underlies the historical development of theory, and some physicists still use it for further development of theory.

Although light and mass differ, they both relate similarly to the invisibility of the virtual field. As for non-detection of radiation by matter, there is the Born rule of quantum physics whereby virtual forms of energy are detectable according to a quantum wave function interpreted as a probability amplitude (as the measure of change) for detecting an atomic particle within some particular time or location. The probability of detection can be extremely slight. Billions of neutrinos, for instance, move through our bodies every second while only a few of them are detected by all of Earth. Neutrinos are thus secondary effects virtually invisible to us for the most part.

Light waves are consistent both with quantum and wave effects in that the numerical constancy of multiplying the parameters m-v-r of the Planck constant is maintained by the change in mass m being nullified by an opposite change in its radius r. Similarly, with light speed being constant, an increase in the relative perception of light energy is nullified by an equal proportion of the light wavelength and the observer's measuring device. Even though the Plank constant applies either to the size of the hydrogen atom or the size of its nucleus, the constancy of more m multiplied by shorter r is consistent with light energy regarding the constancy of its frequency f multiplied by its wavelength λ. The difference is that light can be of any frequency and wavelength whereas atoms are only of particular quantities and sizes in relation to complex structure formation from electron to proton.

Assuming light and matter exist as two different forms of energy that convert from one form to another, further assume matter results from the superposing of light waves that are limited in form to particular spacetime conditions. For explaining this limitation consistent with the relative motion of mass as well as its quantum state, assume equilibrium quantum states of electromagnetic energy become packets of standing waves of mass-energy from which relative motion of the standing waves is caused by further interaction with the virtual field of energy. If a packet of standing waves reflects in one direction more frequent waves than waves reflected in the opposite direction, it then obtains relative motion in the opposite direction in order to balance out the frequency of reflection.

This superimposing condition is consistent with the Doppler effect and covariance of special relativity. Consider a spaceship moving freely in gravitational free space. In the viewpoint of an observer relatively at rest perceiving the ship as relatively in motion, light emitted within the ship in opposite directions of relative motion is of less frequency if from the direction of the ship's relative motion than of the same direction, but they are perceived the same by an observer relatively at rest inside the ship. An object relatively at rest in the ship between the two sources of the light also reflects light back with the same effect of being relatively at rest.

Gravity is also explainable according to this condition of relative motion. How, for instance, can a graviton change the momentum of mass but not itself change in momentum? If emitted radiation results in a vacuum effect, how is it possible for the momentum of mass refilling the vacuum to differ from the momentum of the emitted radiation?

It was previously explained according to its gradual recycling effect. If matter is contained by being in an equilibrium state with general forces of nature, then a particle of gravity propagating through the same equilibrium state of existence can merely pass its momentum on through matter as wave motion of the medium the same as one particle passing on its momentum to another particle. The emission of the gravitational particle then creates vacuum effect consistent with the inverse square law of attraction if there is a source of replacement acting inward, which further explains how gravity continually creates as a particular form of energy without diminishing its source of creation. It is simply according to a recycling process.

The interaction is not totally elastic. If two particles collide head on, their momenta in opposite directions need not be maintained if they are converted to another form of energy such as heat. Gravity is thus being created and gradually recycled back by a slight inelastic amount resupplying the positive force of the medium that it propagates through.

It is the recycling distance that distinguishes gravity as less energetic than that of electromagnetic force. The gradual forming of repulsive force and more containment of it allows less energetic but more continuance change per action, whereas stronger containments of various interactions require more energy for change.

Gravity and Electric Charge

Gravity is only one part of the continuous change of matter. Another part is that of electrostatic charge. What remains to be explained is how gravity and electrostatic charge differ.

Consider electrons as part of an energy field surrounding the nucleus of the atom. The quantum electron effects are such that their electrostatic forces repel each other except for their equilibrium state of quantum existence. Protons within the nucleus also tend to repel each other except for their state of quantum existence. However, although the attraction between electrons and protons is of mutual electrostatic force, the wave action of the opposite forces are themselves such that their interaction results in their transformation into another virtual energy state similar to that of the gravitational one. There is also a recycling process similar to the gravitational one, but of a greater magnitude of both change and restoration.

Gravity is one particular condition of the equilibrium state that applies to all mass, whereas equilibrium states regarding electrostatics differ in various ways.

Gravity and electric charge differ even though they both result from a conversion of particles and virtual particles into other particles and virtual particles. The conversion of particles into slightly detectable gravitons occurs from interaction between all mass. Although Aspden derived the ratio of the gravitational constant from the interaction between protons, it could more generally apply in proportion to the whole mass of the atom. In contrast is the interaction of electric charge combining in different ways for having various positive and negative difference.

Consider the conversion of particles by electrons and protons results in different fields of containment. Both conversions result in fields of positive force for electrons to repel other electrons and protons to repel other protons, but the two different fields overlap for nullification of effect. They themselves convert virtual particles into other virtual particles for recycling effect. The nullification of positive charge thus constitutes a recycling process and a vacuum state of attraction.

It is also possible that the nullification of positive and negative charge could result in the existence of either stronger or weaker gravitational fields. Gravity is thus only one particular aspect of many electrostatic states of equilibrium. Even though the inverse square law applies to both gravity and electrostatic charge, they differ in that gravity does not appear to be countered by antigravity, whereas electrostatic charge results from various differences in magnitudes of opposite charge. As for an example of electrostatic differences, a drift velocity similar to the Clausius mean-free-path principle allows slower action to exist separate from the electromagnetic field for electrostatic action to occur similar to Newton's inverse square law of gravity.

Generally, a weaker state of equilibrium can emerge from a combination of two positive charge particles and one negative charge particle, or from two negative charge particles and one positive charge particle. With positive charge and negative charge maintaining overall balance, new states of equilibrium emerge of less charge per mass. At the proton-neutron level, it becomes a minimum amount of charge. The combining after that is of more mass density and amount of mass maintaining the gravitational equilibrium state of accumulative energy.

Action can either be quantized or continuous. It is quantized for separation of particles. It is continuous regarding non-separation of particles. Gravity being consistent with relative motion is of the latter.

As for the former, there is reciprocal interaction regarding the quantum state of existence. The electron part of the atom is of a longer radius than of the atomic nucleus, but the nucleus has more mass in ratio to less radius. Their resistance to light passing through thus equates to being of equal but opposite charge, as by rotating in the same direction. Deviations can occur of various interactions not resulting in quantum separations.

ABOUT THE AUTHOR

I was born at Childress, Texas in 1943. My dad was a superior athletic, being nominated as a state all-star quarterback during his sophomore year in high school. After serving in World War II, he with me and my mother and older sister moved to Oregon. He and my mother both worked to pay rent and other expenses, and they visited social clubs serving alcohol after work. I and my sisters grew up learning by doing. We moved from one house to another, attending one school for only one or two years.

I enjoyed playing outdoors. I daydreamed while in class, as it seemed to me a prison. I excelled in math, but other than that I was unsure if I was learning what I was being taught, not knowing how other students were learning.

While growing up, I consumed a lot of salt and sugar, even adding more sugar already contained in cereals and other food. Along with countless illnesses, such as measles, chickenpox and so forth, I had less stamina. Later on, I became too hyper from drinking coffee. Although its addictive effect lasted a couple of years, I was able to give it up by mere determination.

I had my tonsils removed when I was young. Later in life, I discovered I had acquired all the symptoms of cystic fibrosis, mainly being unable to gain weight after high school and having no upper fat in my body for my sweat glands to have less water for sweating. More saliva has been produced instead that became toxic regarding allergies. I survived by burping out gas, but as my immune system decreased with age, I later discovered I needed to either spit out saliva or drink a lot less water. Certain vegetables and fruits also contributed to the production of more saliva. Although tomatoes are healthy for the skin and bones, my saliva glands produced more saliva after eating the tomatoes. The outside skins of tomatoes and grapes are their healthy contribution io human skin . Raisins instead of grapes is now what I consider healthy nutrition.

While in my seventies, my nostrils became continually plugged up with snot. A tiny bit of Saline dropped in my nose helped, and I only used a tiny amount as such. Even though its main component is salt water, it also contains highly toxic disinfectants. A small bottle of it generally lasts for a couple years or more.

Another remedy was discovered when my sister lit a vanilla candle. My nostrils suddenly cleared up. However, the resulting effect was not consistent. I gave up on it. However, during early spring when the pollen count suddenly became high, and the passage way to my lungs plugged up, I sipped a tiny amount of Vanilla Extract. My passage way to the lungs was soon cleared. I began sipping Vanilla Extract just enough to counter the amount of pollen during spring and early summer, but I then discovered that drinking more regular cow's milk helped prevent my saliva glands from producing too much saliva. However, the sodium in milk needs to be counterbalanced with potassium. I cook my oats in milk with an added banana. However, although I began breathing better by drinking more milk, I began feeling thirsty for water. I drank a new 7-up loaded with potassium and of zero sugar. The balance worked. I now feel, at age 79, as healthy as ever remember being.

My double cousin, Donna Ticer, who had Huntington's Korea, introduced me to nutrition. Although it did not save her, it benefited me. In my mid-seventies, for instance, I became frustrated of not remembering. I researched nutrition on the internet and came up with pumpkin seeds. After three weeks consuming only a tablespoon per day, I noticed improvement. My memory is nearly back to where it was before. The brain does a lot and needs a lot of nutrition. What works for me might not work for you, but it might be another possibility to consider.

I wondered what vitamins are contained of. I finally found out on the internet that they differ in geometrical constructures that are measured according to effect. Balance is needed. Too much of one shape nullifies its benefit. It depends on what is in and out of the body. What works for me might not work for you.

After serving four years in the United States Air Force, I decided to enroll in a few classes. I missed a question in a calculus class. Although I had the right answer, I solved it by means of geometry instead of calculus. I had not read my reading assignment the night before.

I became interested in physics. I enrolled in a special relativity class and found it to be internally consistent, but I later did not find it, at the time, to be consistent with general relativity.

I worked mostly as a cannery worker until retiring at the end of 1999. I decided to self-educate myself by means of writing and research. Fortunately, experts of countless different categories provided free and countless information on the internet.